Shark Drunk

Shark Drunk

*The Art of Catching a Large Shark from a
Tiny Rubber Dinghy in a Big Ocean*

Morten Strøksnes

*Translated from the Norwegian
by Tiina Nunnally*

ALFRED A. KNOPF

NEW YORK

2017

THIS IS A BORZOI BOOK
PUBLISHED BY ALFRED A. KNOPF

English translation copyright © 2017 by Tiina Nunnally

All rights reserved. Published in the United States by Alfred A. Knopf, a division
of Penguin Random House LLC, New York, and distributed in Canada by
Random House of Canada, a division of Penguin Random House Canada Limited,
Toronto. Originally published in Norway as *Havboka,* by Forlaget Oktober AS,
Oslo, in 2015. Copyright © 2015 by Morten A. Strøksnes and Forlaget Oktober.
Published by agreement with Copenhagen Literary Agency, Copenhagen.

This translation has been published with the financial support of NORLA.

www.aaknopf.com

Knopf, Borzoi Books, and the colophon are registered
trademarks of Penguin Random House LLC.

Library of Congress Cataloging-in-Publication Data

Names: Strøksnes, Morten Andreas, 1965– author. | Nunnally, Tiina, 1952– translator.
Title: Shark drunk : the art of catching a large shark from a tiny rubber
dinghy in a big ocean / Morten Strøksnes ; translated by Tiina Nunnally.
Other titles: Havboka. English
Description: First edition. | New York : Alfred A. Knopf, 2017.
Identifiers: LCCN 2016044185 (print) | LCCN 2016048634 (ebook) |
ISBN 9780451493484 (hardcover) | ISBN 9780451493491 (ebook) |
ISBN 9781524711238 (open market)
Subjects: LCSH: Shark fishing—Norway. | Greenland shark—Norway. |
Fishing—Norway. | Sailing—Norway.
Classification: LCC SH691.S4 S7713 2017 (print) | LCC SH691.S4 (ebook) |
DDC 338.3/727309481—dc23
LC record available at https://lccn.loc.gov/2016044185

Jacket design by Oliver Munday
Interior illustrations by Egil Haraldsen

Manufactured in the United States of America

First American Edition

Have you journeyed to the springs of the sea
or walked in the recesses of the deep?

—Job 38:16

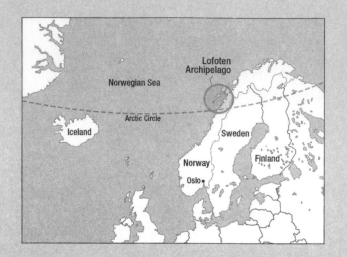

Lofoten Archipelago

Norwegian Sea

Arctic Circle

Iceland

Sweden

Norway

Oslo

Finland

Mockstraumen

Røst

N

W—E

S

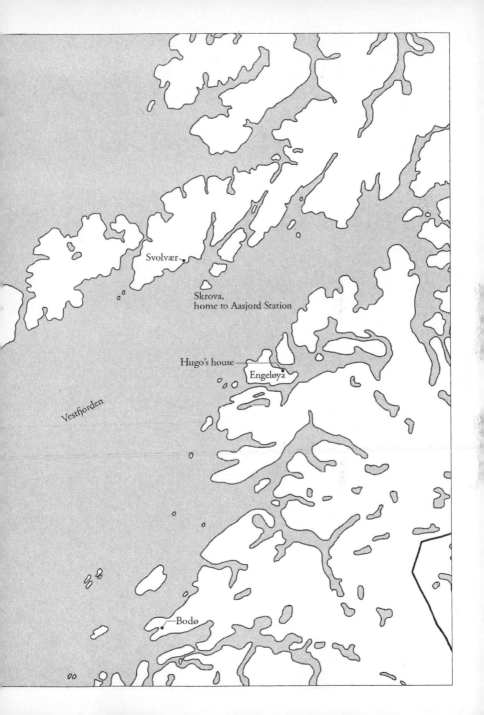

Svolvær

Skrova,
home to Aasjord Station

Hugo's house

Engeløya

Vestfjorden

Bodø

Summer

I

Three and a half billion years. That's the time it took from the moment the first primitive life-forms developed in the sea until Hugo Aasjord phoned me one Saturday night in July.

"Have you seen the weather report for next week?" he asked.

We'd been waiting a long time for a very specific forecast. Not sunshine or heat, not even the absence of rain. What we needed was the least possible wind in the seas between Bodø and Lofoten—or, to be more precise, in Vestfjorden, the "western fjord." If you need calm waters in Vestfjorden, you better not be in a rush. Its tumultuous waters are notoriously temperamental. Even the slightest gust of wind from the west, south, or north can create significant waves.

I'd checked the weather reports for weeks. The forecast was always for strong or gale-force winds. Never just a gentle breeze or light air, which would give us the calm seas we needed. Eventually I'd pretty much given up keeping track and surrendered to Oslo's lazy summertime rhythm of hot days and bright nights.

I was at a lively dinner party when my phone rang. When I saw the call was from Hugo—a man who hates phones and only rings to deliver important messages—I knew that our long wait had come to an end. We were finally going to try to catch the big fish.

"I'll buy a plane ticket tomorrow and arrive in Bodø Monday afternoon," I told him.

"Good. See you." *Click.*

—

On the plane to Bodø I fixed my gaze on the land below. Through the oval window I could see mountains, forests, and plains, which I imagined as a raised seabed. A couple of billion years ago the entire earth was covered with water, except maybe for a few small islands here and there. Even today, the ocean still makes up more than 70 percent of the earth's surface. It has been said that our planet's name shouldn't be Earth. Instead, it would be more appropriate to call it Ocean.

When we reached Helgeland, the land opened into Norway's majestic fjords, with swelling seas to the west, until finally the division between sky and water dissolved and the horizon became a shining gray color reminiscent of bird feathers.

Every time I leave Oslo and travel north, I have the same sense of escape . . . escape from the inland and its anthills, spruce trees, rivers, freshwater lakes, and gurgling marshes. Goodbye and farewell, I'm going out to the sea, which is free and endless, rhythmic and swaying like the old sea chanteys sung across the oceans of the world on ships traveling to classic harbors like Marseille, Liverpool, Singapore, and Montevideo, while the deckhands hauled on the lines to set, trim, or reef the sails.

Sailors who have gone ashore can seem like restless visitors. They may never go to sea again, but judging by their talk and gestures, it's as if they're merely temporary visitors on dry land. They never lose their longing for the water. Yet the sea, which is calling them, has to settle for hopeful, if uncertain, replies.

My great-great-grandfather must have felt the sea's mysterious pull when he left inland Sweden and began walking west. Through valleys and over mountains, he traveled like a salmon along the great rivers, first against the stream, then with the stream until he made it all the way to the sea.

As the story goes, he gave no reason for the journey other than that he had to see the ocean with his own eyes. It's unlikely he had any plans to go back to where he'd come from. Maybe he couldn't stand the thought of spending the rest of his life stooping over barren patches of land in a Swedish mountain village. Clearly he was an impulsive man, a dreamer with strong legs. He wandered all the way to the Norwegian coast, started a family, and then joined the crew of a freighter. A couple of years later, as fate would have it, the ship sank somewhere in the Pacific. Everyone on board drowned. As if the man had come from the depths of the sea and needed to return. As if that was where he belonged, and he'd always known it. At least that's how I think of him.

It was the sea that gave birth to the poetry of Arthur Rimbaud. The sea inspired his expansive language, which carried both him and poetry into the modern era with *"Le bateau ivre"* ("The Drunken Boat") from 1871. The poem is told from the perspective of an old freighter that wants to experience the freedom of the sea and hurtles recklessly down a great river until it reaches the coast and enters open waters, only to encounter a violent storm and sink to the bottom. There it becomes one with the ocean:

> *And from then on I bathed in the Poem*
> *Of the Sea, infused with stars and lactescent,*
> *Devouring the azure verses; where, like a pale elated*
> *Piece of flotsam, a pensive drowned figure sometimes sinks.*[1]

From the airplane I tried to reconstruct more of "The Drunken Boat" from memory. I know the swells attack the skerries like frenzied herds of cattle. And on the ocean floor a Leviathan rots among swaying clumps of kelp, which reel the drunken boat in close, wrapping it in their tentacles. Above the dark abyss of the maelstrom, the boat hears the mating calls of the sperm whale. It sees dead-drunk shipwrecks swarming with sea lice and hideous snakes, golden singing fish, electric crescent moons, and black seahorses—things that people only *imagine* they've seen . . .

The boat is pummeled by the sights, experiencing the frightening and liberating power of the sea, the constant surge and spray, until it becomes languidly, numbingly sated. Then the boat begins to yearn for the dark and quiet river pools of its childhood.

Rimbaud had never seen the ocean when he wrote this poem at the age of sixteen.

2

Hugo Aasjord lives on the island of Engeløya in the municipality of Steigen. To get there from Bodø you have to take the northbound catamaran ferry, traveling between islands and small, weather-beaten communities that cling like barnacles to the rocky coast. After about two hours the ferry docks at Bogøy, a small village with a bridge that crosses over to Engeløya.

Engeløya is like a microcosm of Norway. Typical Norwegian fjord country on the side facing the mainland; archipelago and white shores on the seaward side. The belt of land farthest south and by the sea consists of fertile farmland. Beyond that is a forested strip where moose and other wildlife live. Finally, valleys and mountains, of which Trohornet (2,116 feet above sea level) is the tallest. There are good reasons why people have lived here for nearly six thousand years. This island, which you can easily bike around in a couple of hours, gave them ample opportunities for fishing, hunting, and farming.

Hugo is standing on the wharf and has good news for me. Apparently we have bait. A Scottish Highland bull was butchered a few days ago, and the remains are out in a field, ready for me to pick up. "That can wait until tomorrow," Hugo says as we drive across the bridge to Engeløya and pull up in front of his house, complete with a tower on top, a gallery in the basement, and an unimpeded view to the west, toward Vestfjorden. The square tower, designed by Hugo, could have served as a stage set for Henrik Ibsen's play *The Master Builder.*

Arriving at Hugo's property, it's easy to imagine that you've entered a pirate's lair. Scattered around the garage are objects that have been plundered from the coast. Lining the path to the gallery, the prow of an old ship and several huge old anchors stand like exhibition pieces or trophies. In the backyard is a propeller that once belonged to an English trawler that went down off the island of Skrova. On the shed hangs a Russian sign that Hugo fished out of the sea. He thought the sign belonged to a Russian ship, but it turned out to be an election poster from a district outside Arkhangelsk. Next to the main shed Hugo has built a couple of other sheds, plus a stable housing two Shetland ponies, named Luna and Veslegloppa. Various boats have always been stored in the main shed or nearby. He'd sold the Plattgatter, a mahogany boat with a flat transom that always looked like it was yearning for the Riviera.

Hugo has never eaten a fish stick in his life. Nor does he have any plans to do so. For dinner we have soup made from freshly picked shoots of stinging nettles and lovage, lentils, and home-made moose sausage along with a couple of glasses of wine. Then we go downstairs to the gallery. Hugo's oil paintings are largely abstracts, but people up in the north have a tendency to regard them as actual landscapes depicting the sea and coastline, meaning, of course, scenes from their own lives. That's easy to understand, since the paintings shimmer with the characteristic light found only near the ocean north of the Arctic Circle, espe-cially in winter. Hugo's style is marked by an easily recognizable arctic blue from the cold, clear days of darkness, which, by the way, are actually not dark at all. The entire spectrum of light exists, although dimmed or imploded. The colors of the sky take on a deep, encapsulated glow, while the northern lights can flare up at any moment like psychedelic improvisations.

Some of the paintings he's working on are of Battery Dietl, on the seaward side of Engeløya. There, on the coast, the Germans built the largest and most expensive fortification in northern Europe during World War II. It housed more than ten thousand people, a combination of German soldiers and Russian prison-ers of war. Battery Dietl became one of northern Norway's big-gest towns, with a movie theater, hospital, barracks, dining halls, and even brothels with women brought from Germany and Poland. Spread out through the area were also radar installations, weather stations, and command centers filled with newly devel-oped technology. The cannon battery was intended to cover all of Vestfjorden. It had a range of twenty-five to thirty miles. Today the bunkers still extend several stories underground. Even though hundreds of Russian prisoners died there as forced laborers, Hugo senses a peacefulness in the secluded area.

In his paintings, Battery Dietl shows up only as a group of cubist shapes.

Hugo's work as an artist has a wide range, to say the least. Several years ago Hugo put on exhibition a cat that had been naturally embalmed. Dying, it had taken refuge inside the wall of an old cowshed down the road from Hugo's house. When it became known that Hugo was going to show the cat at the Biennale in Florence, he was asked by the local newspaper, *Avisa Nordland,* "Is a dead cat art?"

Hugo grew up on both sides of Vestfjorden. He has always lived near the sea and also spent a considerable part of his life on boats. Only once did he venture inland for any length of time, when he went to Münster, Germany, to study art. He was the youngest student ever admitted to the city's renowned art school. Back then, many wounded veterans of World War II wandered the streets—men who were disfigured, on crutches, missing an arm, or sitting in wheelchairs. His fellow students were young radical Germans who vocally expressed their criticism of the Vietnam War, while the Second World War remained off-limits. Hugo would sometimes take the train north to Hamburg, because along the way the consistency of the air changed, becoming rawer, with a faint whiff of the sea.

After graduation, Hugo returned to Norway with certificates confirming that he had mastered the classical techniques of painting, graphic art, and sculpture. He also brought with him a different kind of baggage. The fact that he was part of the radical German student milieu in the 1970s still clings to him in a vague sort of way. It has nothing to do with politics, because Hugo has never been particularly radical in that sense. Nor does it have to do with personal style, in spite of his round glasses, mustache, and long black hair. It has more to do with an unconventional attitude about how things should be done and how life should be lived. Münster also left him with a nefarious addiction: watching reruns of the German TV crime show *Derrick* every day at five o'clock. And God help anyone who disturbs him.

After Hugo shows me his new paintings, we go up to the attic room. From there we have a view of Engeløya and its lush terrain. It's a mild summer night. Dew has settled on the grass and the black fields to the south, and a blanket of silence lies over the slumbering land. Even a whisper can carry a long way.

All around us is a bountiful deciduous forest of birch, rowan, willow, and aspen. I go through the open door of the balcony at the prow of the house, which resembles a ship's bridge. It's far from quiet out there. The forest is shaggy with pollen and dripping with chlorophyll. A backdrop of birdsong opens up. I hear snipes, curlews, and woodcocks. My ears need a little time to distinguish one from another. The black grouse clucks, the thrush chatters, the cuckoo sings *coo-coo*. Finches, sparrows, and titmice chirp. Curlews often make a melancholy and lonely whistling sound, but they can shift tempo at any moment, sounding then like a good-natured machine gun. Let's call it friendly fire. Somewhere out there, one bird makes a dry sound, like a coin clacking against a table.

A short-eared owl comes flying past, swooping low. Its long wings flap unsteadily. In the distance, the fjord is smooth and white. The snow hasn't yet melted from the island's black mountaintops, which are high enough that over the years three fighter planes have slammed into them. Two Starfighters crashed in the early 1970s, and in 1999 a German Tornado came down after the two pilots had ejected. Both were picked up by small boats trolling in the waters of Skagstadsund.

The birdlife says a lot about the difference between the islands of Engeløya and Skrova, which is on the other side of Vestfjorden. Engeløya is an agricultural community. Skrova is a fishing village, where everything, including the way of thinking, is different. There are only seabirds on Skrova. The birds of the forests in Engeløya can sing with enchanting beauty, whereas the seabirds around Skrova often have hoarse, croaking voices. But some of those seabirds can dive down to 650 feet, practically flying through the water and constantly changing direction as they close in on panicking shoals of herring or sprat.

Offshore of Skrova, the sea level dramatically plunges a thousand feet. There, on the coast, Hugo and his wife, Mette, are in the process of fixing up Aasjord Station, an old fishing outpost and cod-liver-oil mill.

As the name suggests, Hugo's family owned Aasjord Station for a couple of decades before it was shuttered and sold in the early 1980s. Now Hugo and Mette have bought it back. The place had fallen into serious disrepair, but they've partially restored it to its former glory, with even bigger plans for the future.

As for Hugo and me, Aasjord Station is going to be the home base for our shark hunt.

—

Back inside, Hugo tells me the story of the rams, which would be a strange tale coming from anyone else, but it's somehow normal for him. I'm not sure why he thought of it now, but he does tend to meander from topic to topic. Once, he adopted a practically newborn ram because the farmer thought there was something wrong with it and was going to put it down. Hugo felt sorry for the ram and took it home. The ram moved into the kitchen, and he and Mette planned to slaughter it in the fall. A few weeks later Hugo was in a shop and happened to run into the same farmer, who casually remarked that it was a shame for the ram to be alone. So the farmer dropped by with another ram of the same rejected category.

Over the following months and years Hugo and his family fed the rams until both were big and strong—and completely unmanageable. It was no longer safe to let them near the kids or the dogs, so Hugo loaded them on board his boat and took them out to an islet where they could stay and graze.

They grew big and strong and forgot how to say thank you. Whenever Hugo passed the islet they would often swim toward his boat, their wet, heavy wool pulling them under, and then Hugo would have to rescue them. One beautiful summer day, Hugo calmly attempted to go ashore. Not sensing any danger, he was halfway out of the boat when one of the rams charged him. Hugo pulls up the sleeve of his sweater to show me a big scar on his upper arm and ends the story with this physical exclamation mark.

A short time later, both rams were slaughtered. The Aasjord family had lost all sympathy for the two creatures. Now, their pelts hang over a rod in the small shed.

—

It was on a night like this one, two years ago, when Hugo first mentioned Greenland sharks. Hugo's father had gone on whale hunts from the age of eight, and he'd seen how the Greenland shark would emerge from the deep and steal huge pieces of blubber from the whales as the crew drove and flensed them alongside the boat.

Once, the crew harpooned a persistent Greenland shark and used the derrick to haul it up by the tail. Even in its half-dead condition, as it hung upside down with a whale harpoon stuck through its back, the shark gorged itself on the fresh whale meat within reach. That Greenland shark took forever to die. It lay there for hours, eyeing the crew as they moved about on deck and spooking even the toughest and most seasoned fisherman.

Another hot summer day while drifting in Vestfjorden aboard the fishing boat *Hurtig,* one of the fishermen decided to cool off by going for a swim and dove into the sea. When a Greenland shark suddenly surfaced just a few yards away, the fisherman scurried back into the boat in a flash—much to the amusement of the rest of the crew.

Stories like this fueled Hugo's imagination, brewing inside of him for forty years. That night, when he talked about the Greenland shark, a gleam appeared in his eyes and his voice took on a special tone. The tales he'd heard as a child had never lost their hold. He had seen most of the fish and animals that live in the ocean, he said, but he'd never seen a Greenland shark. I hadn't either. Hugo didn't have to work very hard to convince me it was time. I took the bait, so to speak. Hook, line, and sinker.

I also grew up near the sea and had gone fishing ever since I was a little boy. Getting a nibble always gave me the feeling that just about anything might come up from the deep. A whole world existed down there, containing countless creatures that I knew nothing about. In books I'd seen pictures of the known species, and that was more than enough. Life in the ocean appeared richer and more exciting than life on land. Strange beings swam around, practically right under our noses, but we couldn't see them and didn't know them. We could only guess at what went on down there.

Ever since, the sea has retained its attraction for me. Much of what we find mysterious or exciting when we're children loses its aura by adolescence. But for me the ocean simply grew bigger, deeper, and more amazing. Maybe an atavism came into play; something had skipped over several generations in my family before I inherited it from my great-great-grandfather, who had ended up at the bottom of the sea so long ago.

And there was something else, something I wasn't fully aware of at the time and may not see clearly even now, except on the periphery of my vision—like when the rotating beam of a lighthouse rips apart the darkness with a swift flash of light.

There were actually a lot of other things I should have been doing when, without hesitation, I replied, "Sure, let's go to sea and catch a Greenland shark."

3

We have mapped the world, and we no longer fill in the blank spaces with strange monsters and fantastical animals conjured from our imagination. But maybe we should, because all life on the planet has not yet been discovered—far from it. Scientists have catalogued just under two million species of animals to date, but biologists estimate that altogether there are about ten million multicelled organisms on our planet.[2] Undoubtedly the greatest discoveries await us in the ocean, where new life-forms are constantly being discovered. In fact, we still know very little even about large creatures that swim near the coast. There could be as many sharks in the ocean as people on earth.[3] And who knows anything about the Greenland sharks swimming in the deep troughs and channels of Vestfjorden, sharks that can grow to twenty-four feet in length and weigh twenty-five hundred pounds? Other than Hugo, of course.

The Greenland shark is prehistoric. Reportedly, it swims at the bottom of deep Norwegian fjords and all the way up to the North Pole. Yet deep-water sharks are usually much smaller than those that live at shallower depths. The Greenland is the major exception. It can grow bigger than a great white and is thus the world's largest flesh-eating shark. (The basking shark and whale shark are bigger, but they eat only plankton.) Marine biologists have recently discovered that the Greenland shark can reach an age of four, maybe even five hundred years. That makes it by far the oldest vertebrate on the planet. The shark we're going to catch could have been swimming slowly around in some dark oceanic abyss well before the *Mayflower* set sail for the new colony of North Virginia, or even a hundred years earlier, when Nicolaus Copernicus concluded that the earth is orbiting the sun. It could be half the age of Methuselah. According to tradition, Methuselah perished in the year of the Flood, and maybe the swelling water was what finally got him. The Greenland shark would have found the altered circumstances on earth very agreeable, considering the unparalleled abundance of food that must have been available.

One more thing: in Norway, people sometimes assume the Greenland shark is related to the porbeagle. But we're talking about two different species. The porbeagle is smaller and has tasty flesh that could potentially be served in a restaurant—if it were not on the endangered list, of course. The Greenland shark is wild and not on any list, but few people would ever want to eat the flesh of its massive body. Its meat is laced with a toxin that, when ingested, produces a feeling of extreme intoxication and can be fatal.

Still, cost what it might, we were going to catch a voracious monster with many hundreds of millions of years of evolution behind it, with potentially fatal poisons in its bloodstream and teeth like that of an oversized steel trap, only a lot more of them.

It was two years ago that Hugo and I first made our decision, and now the summer night sky takes on the orange tinge of caviar. We're sitting here exchanging Greenland shark news, because we've both found out a thing or two since we last met. Most sources in print describe it as slow and sluggish. Unbelievably, the swiftest kinds of shark can reach a top speed of almost forty-five miles per hour. Hugo doesn't agree that the Greenland shark is so much slower.

"Then how do they explain that the remains of polar bears and the fastest fish in the sea, including halibut and mature salmon, have been found in the stomachs of Greenland sharks? *How slow could it be?*" asks Hugo.

"Most Greenland sharks carry a parasite that attacks the cornea and makes them partially blind. Some are depicted with what looks like finger-long worms hanging from their eyeballs. One theory is that the prey is hypnotized by the shark's eyes, which are a luminous green in the dark," I say, pleased that I can tell Hugo something about the sea that he might not already know.

My joy is short-lived. Hugo is not impressed.

"If that's true, then how is the shark able to bring down reindeer in Alaska? And how does it catch seabirds? Are you saying it hypnotizes them too?"

Hugo launches into a brief lecture on the sensory organs of the Greenland shark. If the shark is blind or partially blind, this would be less of a handicap than you would imagine, because it's so dark down there in the deep anyway. But the Greenland shark has a secret electromagnetic weapon. Like many other sharks, Greenland sharks are equipped with so-called ampullae of Lorenzini. Through these jelly-filled, half-inch-long ampullae, it can sense changes in electric fields down to a few billionths of a volt. That's probably how it detects prey buried in the sand and how it manages to sneak up on seals lying or sleeping on the ocean floor before attacking.

I look at him, trying not to show that this is new to me.

"Didn't you know that seals sleep on the ocean floor?" he says, a bit gleefully, before continuing his lecture.

"Maybe the Greenland shark uses these properties to catch animals that are much faster, or maybe it finds fish that are injured or weak or buried in the sandy bottom. Maybe it usually moves slowly and soundlessly, perfectly camouflaged, snapping up its prey . . ."

I can tell that he's about to get to the point.

"But I'm pretty sure it's capable of increasing speed in sudden lunges. That's the only logical explanation," he concludes.

We haven't yet discussed certain details. For instance, what do we do if we actually bring a Greenland shark up to the surface? I suggest that maybe we can try to tie a rope around the base of its tail and haul the shark up backward so that it passes out. Sharks have to keep swimming in order to take in oxygen. The same is true of mackerel.

Hugo shakes his head. He thinks we would risk the shark sinking. Maybe instead we should try to steer it toward land, the way the Inuit do. The weak link in this plan is that we'd have to persuade the shark to swim in the direction we want to go. The Inuit use two small kayaks on either side, steering the Greenland shark between them, while we have only one boat. By the way, the Inuit traditionally regard the Greenland shark as one of the animals that help shamans.

"Maybe we could pull it up onto an islet—if we can get the islet between us and the shark," I say.

Hugo blithely ignores my suggestion, presumably because it's so stupid.

"What if we drag the shark ashore? If we have time to wrap the rope around a tree, we can then move off in the opposite direction and pull the shark all the way up onto land," I propose.

"A little better, but I've been thinking about this, and I know what we need to do. When the Greenland shark comes up to the surface, we stick another shark hook into it and tie it to a float with a short line. Then we can figure out the rest."

If we manage to get the shark, either backward or forward, to one of the docks or beaches around Skrova, Hugo is interested in the liver. He can extract a barrel of oil from the shark's liver and use it to make paint. Then we'll paint Aasjord Station as part of its renovation. Hugo is pondering various art projects for which he could use the shark.

After going on like this for a couple of hours, we've run out of juice. This is not the time of the midnight sun, but it's still bright daylight. I sit outside on the porch to stare at nature. It's truly a pliant night, with almost no breeze. From the waters of the sound comes a faint hint of salt and rotting seaweed. All our gear is ready and waiting on Skrova, at Aasjord Station. We have chains and thirteen hundred feet of the best quality nylon line. We have shark hooks eight inches long and made of stainless steel, and weights to make the line sink. We have two big floats to absorb the pull if the shark bites, so it will wear itself out and, if necessary, we can keep it a safe distance away.

The only thing we're missing is bait. The time has come for me to gather up the remains of the Scottish Highland bull lying somewhere out in a field. Hugo doesn't have the stomach for it. After a botched operation, he often feels the urge to retch, even though it's now physically impossible for him to vomit.

Luckily that's something I can still do.

4

Life cannot exist without death, and the cycle of life is what keeps the planet in harmony. At least that's my philosophical solace while, early the next afternoon, I trudge alone through the woods, following vague directions about where to find the sure-to-be-rotting carcass of a Scottish bull.

Scottish Highland cattle are a primitive and hardy race that spend the whole winter outdoors and look something like musk oxen with long bangs. They are herd animals with a strict hierarchy. It's best not to get too close when they are calving because these animals have retained all their natural instincts. With their long, sharp horns and enormous strength, these ancient creatures can do much more damage than an aggressive ram. Highland cattle often scare the living daylights out of berry pickers.

A farmer had been raising these animals for a couple of years. Or rather, he'd left them to graze in the forest while he worked on an oil platform in the North Sea. The first time he slaughtered one of them, he used a humane killer, which drives a bolt into the animal's forehead and kills ordinary large cattle instantly. But the forehead of the Scottish bull is two and a half inches thick, and it turned out that the bolt merely knocked the beast unconscious. It looked dead, but as soon as the farmer had severed the main artery, the bull got up and began running around, panic-stricken, blood gushing out over the farmer and his kids, who barely managed to escape to safety.

The bull that was now our bait had to be shot several times with a .308 rifle, which can kill a moose at a distance of half a mile. Only after the third shot did the bull finally keel over.

But where is the carcass?

I follow the directions and reach a field. According to what I've been told, the remains of the bull are lying among the trees on the far side of the field. It's a summer day, one of those warm and sunny days that you rarely get this far north. Songbirds are chittering as if they've drunk champagne for lunch; bumblebees are lazily buzzing around the flowers. I see red clover, oxeye daisies, geraniums, and multitudes of that puffy yellow flower that has so many names: bird's-foot trefoil, bacon and eggs, Dutchman's clogs, lady's slipper, granny's toenails, and devil's fingers.

This flower has a particular fragrance that has locally prompted nicknames of a highly profane nature: shit-stink flower, Satan's diarrhea, and perhaps the least appealing name ever ascribed to a flower: ass-wiper gut grass.

At any rate, it would have been a perfect day for a picnic on Engeløya, except for the whole carcass thing.

Not far from where I'm searching for the carcass, there is an old sacrificial altar called a *horg*. Through Hugo I've developed a certain interest in the hollowed stone, since he has put it in one of his paintings. Povl Simonsen at the University of Tromsø is one of the few people who has written anything about the *horg*. He maintains that there are only two sacrificial stones of this type in northern Norway. One is on the island of Sørøya in Vest Finnmark; the other is at Sandvågan on Engeløya. Simonsen dates the stone to somewhere between 1000 BC and AD 1000.

It's astonishing how imprecise this is. Simonsen says the stone could be either from the late Bronze Age or the late Iron Age. The explanatory text on the sign recently posted next to the stone by the Directorate for Cultural Heritage is not doing any better. It says that the stone is from the period between 1500 BC to AD 1000. In other words, the stone might be 3,500 years old, or maybe only 1,000 years old, which means that no one has any idea who used it or when or how. It's rather like reading in the newspaper that the new world record in the hundred meters was under one hour, and that it was set by a man or a woman between the age of one and a hundred.

Because of the hollows in the stone, it was likely used for sacrifices. The hollows might have been meant to collect the blood or fat from humans or animals. The stone faces west, which prompts speculation that it might have had something to do with sun worship. Maybe the people sacrificed virgins, or merely domestic animals, or maybe just milk, butter, and grain. Maybe they held sacrificial celebrations once a year. That was one way to bind people to a specific group. Everyone participated, and there was guaranteed to be music, dancing, food, and intoxicating drink. In my mind a certain thirst for blood was also involved. A way of remembering or reenacting the violence that had brought their forefathers together in groups.[4]

So I'm wandering around speculating about animals and sacrifices when a little gust of wind sweeps across the ground in my direction. Judging by the smell, I'm on the right track. The stench makes me gag, which in turn brings tears to my eyes, and I stumble over a big tussock and land in cow shit. After a night of bingeing on red wine with Hugo, I'm not really prepared for this. Halfway across the field I can hear the flies. Hugo has sent along with me what I thought was a gas mask, but it turns out to be a dust mask, which is totally worthless in fending off the stink of death. In our part of the world most people have forgotten what death smells like. The stench begins spreading almost instantly after the body dies, but it only gets really bad after three days, when the bacteria inside the stomach start feeding upward to consume their dead host. During the process, waste gases and extremely toxic fluids are created. Our sensory organs explicitly warn us to stay as far away as possible from such poisonous substances. Not to seek them out as I'm now doing.

A renowned evolutionary biologist once described us humans, no matter how high standing and cultivated we may be, as a thirty-two-foot-long canal through which food passes.

Everything else we have acquired through evolution—the brain, glands, organs, muscles, skeleton, and so on—is extra equipment built around this canal.

There's not much use in reducing the human being to such a basic function. But the earth's most widespread life-form, with the exception of microorganisms, is a canal surrounded by a muscle: worms and maggots. Hardly any other creatures have colonized the earth more efficiently, and nowhere are there more of them than on the floor of the ocean. A dead whale carcass becomes the home for millions of worms.

Tens of thousands of whales die every year. They are not buried in mythical whale cemeteries to the sound of mournful whale songs accompanied by some sonorous hydraulic sea organ of the deep. Some drift ashore, but most sink to the bottom. The smell attracts carrion scavengers from far and near, and they establish what is called a whale fall community. A slow explosion of life occurs when colonies of various types of parasites establish themselves. They can keep going for decades before the whale skeleton is stripped clean. And even the bones become food. A special type of worm that looks like a tiny red palm tree will attack the skeleton. And even that is not the last meal these carcasses provide. Soon the bacteria take over. They turn toxic sulfides into nutrient-rich sulfates. This process alone will provide nourishment for four hundred different species, including bivalves. And after everything has been consumed, all of these species drift onward, surviving on standby mode as they search for the next oasis. This is something we know a lot about, since scientists have lowered dead, stranded whales down into the deep to study exactly what happens.[5]

I find the bull carcass and start to pack up its robust bones and rotting intestines in sacks. Tears run from my eyes, flies buzz around my ears, and the sun shines just like it does on a lovely day. It suddenly strikes me that Hugo should have done this job. Why did I buy his argument? His being unable to throw up shouldn't have excluded him. It makes him superbly qualified.

5

Two hours later we're at the harbor in Bogøy, ready to cross Vestfjorden in Hugo's RIB (rigid inflatable boat). It's a French-built boat made by Bombard, which makes me think of it as a potent weapon of destruction. In reality, it's only a small boat, made of rubber and filled with air.

We load the sacks and the rest of the gear on board, inflate the pontoons using the mechanical foot pump, and set out across Flaggsundet at a speed of thirty-seven knots, made possible by a recently overhauled 115-horsepower Suzuki motor. The RIB is unlike any of the other boats Hugo has owned. It can reach a speed of forty-three knots, or fifty miles per hour. Since it hardly has a keel and is filled with air, it floats on the surface instead of in the water. I can see why Hugo loves his RIB. It can walk on water.

The history of Hugo's family follows the boats they have owned. For generations the Aasjords have been involved with various types of fishing and hunting, including whale hunts. Hugo's great-grandfather Norman Johan Aasjord—originally a cantor, cabinetmaker, and teacher—was a pioneer in the development of Norwegian fisheries. He started out on his own, and after spending time as a fish buyer in Finnmark, he took over a fishing station in Helnessund, in Steigen, south of Engeløya. On the mountain high above the station, he built an artificial pond, which froze solid in the winter. All summer long, ice could be sent down to the station on a wooden chute, which made it possible to export fresh fish to Europe.

Hugo grew up in Helnessund, where he ran in and out of the family fishing station all year round. In the winter the kids played in the fish-drying loft. The call of the sea starts early. Even the oldest sailors probably started going out onto the water as young as eight. When Hugo was only ten, he and his pals would often stay out all night in small boats to fish or spear wolffish with a *pik*—a weighted harpoon that you drop from the boat. Since light refracts in the water, it's an art to calculate correctly where to aim when you see a wolffish or flounder on the bottom. Another method is to drop a hook and line over the side of the boat, wait until you actually *see* the fish getting very close, and then yank the line up at exactly the right moment. Both techniques require training and precision, but when a boy masters them, he feels like he's king of the universe.

The big blue wolffish are so aggressive that they'll come back if you miss, while the small brown ones realize it's best to take off. Once when Hugo and his brother and father were out spearing wolffish, they caught a big one that wriggled free at the surface. All three leaned over the gunwale to look for the wolffish on the sandy bottom. It seemed to have disappeared without a trace. Then they heard the keel of the wooden boat start to creak.

Norman's son, who was Hugo's great-uncle Hagbart (not to be confused with Hugo's father, Hagbart, or Hugo's four-year-old grandson, Hagbart), was a legendary innovator in the district.

He instituted new methods and started catching types of fish that no one had previously valued.

Great-Uncle Hagbart's whaling career began in a roundabout way. He was fishing for halibut off the west coast of Canada and Alaska when an American friend, who was a harpoon maker, introduced him to the whaling world. Several years later, when Hagbart returned to Bodø, he had a harpoon made and then borrowed an old cannon that had been used to shoot basking sharks—the plankton-eating shark that is the second-biggest fish in the sea after the whale shark. Basking sharks swim around with their jaws wide open to filter the water. It makes for a slow and peaceful way of feeding, but the basking shark looks aggressive, if not totally insane.

Basking sharks were in demand for their livers, but it could be dangerous to get too close. If the boat got between the sun and the shark so that it saw its shadow, the fish would use its tail to strike. The blow could lift the boat into the air, make it capsize, or even crush it. So hunting for basking sharks required a great deal of care and accuracy. Many used handheld harpoons, which had to be thrown the instant the tail was right next to the boat. Then the shark would strike in the opposite direction as the harpoon bored into it.

People laughed when Hagbart said he was going to begin whaling, but after some trial and error, he was bringing in up to thirty minke whales a week. Three boats were rigged and equipped for the job. That was how commercial whaling came to Steigen and Vestfjorden. The small island of Skrova in Lofoten, where Hugo and I are now headed, eventually became its epicenter. Today it's still one of very few places in Norway that serves as a landing center for whales.

One time Hagbart and two colleagues harpooned a huge fin whale. The fin whale can get almost as big as the blue whale, which is the largest animal on earth. Its sleek, cigar-shaped body also makes it faster than most other whales. The fin whale dragged Hagbart's small boat for dozens of miles, straight across outer Vestfjorden, all the way to what's known as the Lofoten Wall, a long series of mountain peaks that from a distance look as if they form a continuous line rising straight out of the ocean.

This story is no exaggeration. In 1870 the Norwegian author Jonas Lie was on board when a fin whale dragged a steamer belonging to the whaling pioneer Svend Foyn across half of Varangerfjorden, some five hundred miles northeast of Vestfjorden. The ship was pulled against the wind, and the steam engine worked hard to put on the brakes, but it did little good. Foyn also raised a jib sail, but the wind ripped it apart. The swells poured over the bow, and the crew wanted to cut the whale loose, but old Foyn merely paced back and forth on deck in a contemplative mood. Jonas Lie writes: "The situation was getting more and more unpleasant; it was as if we'd harpooned the god of the sea instead of a whale, so incessantly and relentlessly did it run. When the hawser finally broke, no doubt many a soul on board breathed a sigh of relief."[6] The experience caused Foyn—who invented the harpoon grenade and thereby increased sixfold the efficiency of whaling ships—to devise a crossbeam with "ears" attached that would stick up vertically in the water. Lowered into the sea, it would significantly enhance the boat's ability to brake.

The Aasjord family has owned fish landing stations, fillet factories, and cod-liver-oil mills, as well as export companies that sold fresh fish, salt fish, dried fish, and dried-and-salted cod, called *klippfisk*. Their boats have formed the hub of all these enterprises. When Hugo talks about his grandparents, his father, his uncles, and old friends, he almost always mentions the boats they've owned. Although he has never shown me photos of his relatives, I've seen plenty of pictures of their boats. I can't even count how many times I've heard the names: the *Hurtig*, *Kvitberg I, Kvitberg II,* and *Kvitberg III*; the *Havgull* and the *Helnessund,* both I and II. Or the *Elida,* an old Plattgatter sloop, meaning a wooden vessel with a flat transom, jib boom, and gaff rigging, which the family firm owned up until the 1930s. There was also a trawler that came to Steigen from Iceland with a big dent in its bow after colliding with a British navy vessel during the cod wars of the 1970s.

Hugo was only eight when the *Kvitberg II* sank, but he talks about the boat as if it were a dear old family member. It was a seventy-four-foot cutter that went down off Stabben, on its way from Bodø to Helnessund. On deck was a cargo of lime, cement, and septic tanks. Outside Karlsøy the wind picked up, and heavy seas caused the cargo to shift. The boat sank almost instantly. Hugo remembers that his uncle Sigmund waded ashore in Helnessund, soaked through and his whole body chalk white. The cargo had dissolved in the water as the boat went down, coating everyone on board.

The *Kvitberg II* is not the only boat belonging to Aasjord & Sons to be wrecked. Just after the new year in 1960, the *Seto* went down off the coast of Møre. It was a trawler that had been converted into one of Norway's largest purse seiners, and it had just made a big haul and taken on board 84,535 gallons of herring. The boat was about to head for shore to deliver the fish when it heeled over, capsized, and sank in seconds. The crew scrambled and were quickly picked up by a nearby boat. The next day *Bergens Tidende* reported, "It was a dejected bunch that arrived in Ålesund in the early hours of Saturday morning on board an assist boat after their own ship, the purse seiner *Seto* from Leines, near Bodø went down in the herring grounds 10 nautical miles west of Runde. The crew failed to salvage any personal belongings. Even their wallets remain on board."[7] Captain Ludvig Åsen thought that a bulkhead in the cargo hold must have burst so that several dozen tons suddenly shifted. If this had happened on their way toward land, with no other boats around, things could have gone very badly for the twenty on board.[8]

After World War I, Hugo's grandfather Svein and his great-uncle Hagbart bought a British minesweeper. It was made of oak so that it wouldn't detonate magnetic mines. Whenever Hugo talks about the *Cargo,* as that boat was called, you can hear the longing in his voice. He almost makes it sound as if life is missing an important element if you don't have access to a British minesweeper made of oak.

On our way out to Flaggsundet we pass the fish farm, and I happen to think about the *Kvitberg I* and what Hugo had told me about the boat. It was a solid vessel, built as an icebreaker in 1912. In 1961, after it had served its purpose, the boat was parked in the intertidal zone of Innersundet in Helnes, where it eventually fell apart and was swallowed up by the sand. There it would normally have stayed until the last beam rotted away.

But Hugo had other plans. In 1998, he had the prow dug up

along with a section of the ship's side. Both pieces were put on display at the premises of the Bodø Art Association. Bjarne Aasjord (1925–2014), who was the last owner of the ship, couldn't really understand what his old boat was doing in an art exhibit after it had been buried for almost forty years. But for the first time in his life, he attended an art opening.

After the show was over, Hugo placed the hull near the salmon farm in Steigen. There it stayed for several years, until it was once again buried on the foreshore, though no one told Hugo. Now he's considering digging it up again, maybe to put it in another show. That ship's hull must be starting to wonder what's going on.

Fishermen frequently talk about their boats as if they were alive. If pressed, they'll admit that of course the boats are inanimate, but deep inside they know this commonly held view is all wrong. It may be because they're so closely connected, and in an emergency the attributes of the boat can spell the difference between life and death. It's essential for the fisherman to know the boat's personality, its quirks, strengths, and weaknesses. Together they can master the sea, provided the boat is handled with respect. Today, of course, speaking about boats in this way is no longer common, except for people like Hugo.

Hugo makes the boats sound amiable, clever, diligent, nice—or difficult, cantankerous, maybe even deceitful. He speaks fondly of most of them. It's true that they may have had their quirks and eccentricities, but if you showed them respect and coaxed forth their secrets, they would prove to be amazing vessels. When Hugo talks about them, it's as if he prefers to emphasize their positive aspects rather than their flaws and weaknesses, in the same way people tend to speak kindly of deceased friends. We all have our limitations, after all.

Ten years ago Hugo owned a Viksund, though he never felt truly comfortable with it. When the wind kicked up and the boat began to roll, sediment would come up from the diesel tank and clog the filter, which could make the motor stall. That can be dangerous in the treacherous waters where he usually ventures, such as the area south of Engeløya, out toward Engelvær—especially if it's dark and you have two young children sleeping in the bow. The Viksund's motor was unreliable, and even though the boat never wrecked, Hugo always speaks of it with a trace of scorn.

And by the way, I too have bad memories of the Viksund. One time the wind was so strong that the tub of a boat began to roll. I got really seasick, and Hugo thought that was the perfect moment to tease me. As I hung over the rail, he put on a concerned expression and said, "It's always been a mystery to me why people get seasick. Do you do it on purpose? I've always been curious to know what it feels like, but it's never happened to me. How exactly would you describe it?"

As far as I can recall, I tried to grab his scarf so I could stuff it into the propeller, but I was too weak. Later he told me that he actually used to get horribly seasick, up until the age of fourteen. He was often so ill that his parents would put him ashore on a bare islet just so he could feel solid ground under his feet.

The RIB races out of Flaggsundet, and Vestfjorden quickly approaches. The sheltered waters are completely calm; the only ripples on the sea are those we're making ourselves. Hugo can just "plow ahead," as he says, at least for the time being.

The conditions almost always change once we come around Engeløya and enter Vestfjorden. It's not really a fjord at all, but rather a moody stretch of sea. Some people call it the Lofoten Pool, which always makes me think of the world's biggest and coldest swimming pool. The place where we'll cross is about seventeen nautical miles as the crow flies. Vestfjorden is one of the areas sailors and fishermen often talk about, in addition to Hustadvika, Stadthavet, Folla, and Lopphavet. In any case, it's one of the largest ship graveyards along the Norwegian coast.

The phenomenon called *storsjøtt* in Norwegian is something else that makes Vestfjorden extra tumultuous. During a full or new moon, when there's an extreme difference between high and low tide, enormous amounts of water are forced into the narrow and deep Tysfjorden. At low tide the huge volume of water surges back, colliding out in Vestfjorden with currents driven in by a southwesterly wind. This creates big swells of seawater and unpredictable currents.

There are skerries all along Vestfjorden that have crushed countless boats into kindling and left behind many widows and fatherless children. If you study the sea charts of these areas, you can tell a great deal just from the names of the shoals that are either barely visible or lie just below the surface. Bikkjekjæften (Dog Jaws), Vargbøen (Wolf Lair), Skitenflesa (Shit Rock), Flåg-skallene (Floe Skulls), Galgeholmen (Gallows Islet), Brakskal-lene (Crashing Skulls). Whenever a storm occurs, the sea rages around these islets and skerries. Most of them may not look like much, but what's visible at the surface might be the peaks of huge submerged mountains, with the whole ocean crashing against them with violent force. Some skerries are visible only under such conditions. They are the most treacherous of all.

In the old days, the fishermen often had to stop and wait for weeks at the old trading post on the island of Grøtøy or in one of the smaller, outlying fishing villages on Vestfjorden until the waters were calm enough to cross. Then they'd end up in debt to the merchant and fishing village owner Gerhard Schøning,[9] who would subsequently have the men in his pocket. In the late 1800s, he would board the steamship *Grøtø* and travel all around the fjords, visiting villages and telling those who owed him money which party they should vote for. The conservative party Høire won some surprising victories among the fishermen and farmers who were weighed down by debt.

The fishing village owners divided up the sea among themselves and blocked anyone but their own fishermen from fishing in their areas, resorting to force if necessary. If there was a lot of fish, the owners would collaborate and demand two fish for the price of one, thereby cheating the fishermen out of half their pay. Feudal conditions existed, and in many ways the fishermen were tenant farmers subject to the rule of the fishing village "lords."[10]

6

After about half an hour we're finally out in Vestfjorden's wide, open waters, inhabited by multitudes no one can count. Ships pass. The Leviathan plays.

It's not like those painful crossings when we zigzag to avoid meeting the sea swells head-on; they can slam against the boat and make your flesh feel like it's being shaken off your bones. Not this time. We can already see details of the Lofoten Wall on the other side, appearing magnified in the warm, clear air.

Parts of the jagged black peaks have been here since the earth's beginning.

Right now the water is as still as liquid white metal, just as Hugo had predicted. It's one of the year's calmest days on Vestfjorden. We scan the mountains of the Lofoten Wall from one end to the other. We can glimpse the village of Lødingen in the northeast, then the peaks and islands of Digermulen, Storemolla, Lillemolla, and Skrova, which hides the town of Svolvær and the approach to the village of Kabelvåg. Continuing west, we see the sharp silhouette of Vågakallen, and the fishing ports of Henningsvær and Stamsund, Out there toward Lofoten Point, cloaked in a sleepy veil of mist, are Nusfjord, Reine, and Å. At the very end is the infamous Moskstraumen, a maelstrom feared by sailors for centuries and described with glee by authors such as Jules Verne and Edgar Allan Poe.

The view of the famed Lofoten Wall has caused many to gasp. When the Norwegian painter Christian Krohg crossed Vestfjorden on a winter day in 1895, he wrote, "Well, it cannot be denied—an impressive sight. The purest of the pure, the coldest of the cold, the most virtuous of all, the grandest imaginable, altars to the god of solitude and chastity's divine virginity. Difficult—how difficult to paint this! To convey the elevation, the grandeur, and nature's inexorable, merciless calm and indifference."[11]

Krohg had no interest in painting Svolvær, the "capital" of Lofoten. In his opinion, the town did not fit with the landscape; it seemed out of place. Its brown color was too abrasive, without a consistent nuance or mood. In any case, it lacked all sense of harmony with the light and with nature.

—

If Krohg had known what exists down in the deep, he might have become the first surrealist painter. On land, life is lived horizontally. Almost everything takes place on the ground, or at most on a level with the tallest trees. Of course birds can fly higher, but even they spend the majority of their time near the ground. The sea, on the other hand, is vertical, an interconnected column of water with an average depth of approximately 12,000 feet. And there is life from top to bottom. The vast majority of living space on earth, so to speak, can be found in the sea.[12] All other landscapes, including the rain forests, pale in comparison.

If we combine what we know about the ocean's depths, from a purely logical point of view we can conclude that everything found on land—all the mountains, ridges, fields, forests, deserts, even the cities and other man-made phenomena—all this could easily fit into the sea. The average elevation on land is only 2,700 feet. Even if we dumped the whole Himalayan range into the deepest part of the ocean, it would make only a big splash before the mountain chain sank and disappeared without a trace. There is so much water in the ocean that if we imagined the entire seafloor rising up to what is now the surface, all the continents would be totally covered under many miles of salt water. Only the tops of the tallest mountains would stick up out of the ocean.

We find ourselves on a surface of dazzling bright sunshine and mirrorlike water. In Lofoten the local word for this is *transtilla* (a term derived from the Norwegian word for cod liver oil), meaning those rare occasions of perfect calm. Up ahead the sea is 1,600 feet deep. We have no idea what is going on beneath the almost white surface membrane. Well, that's not quite true. Living in the kelp right below us are coalfish, haddock, cod, pollock, and many other species. Under the kelp forests, at a depth of 500 to 650 feet, almost all light is absorbed by the water, no matter how clear and pure it might be. A distant grayish glow, like the light from a dying old TV, is all that's visible. Photosynthesis ceases; it's over and out for all plants. At these depths and farther down, remarkable species live in a darkness that is constantly patrolled by the Greenland shark.

What goes on in the vast depths of the ocean has always been a mystery to us. Only in the past hundred and fifty years or so have we known anything at all about it. During this period our understanding has made halting progress, with new knowledge completely replacing most of the old. In 1841, after an expedition to the Aegean Sea, the leading British naturalist Edward Forbes concluded that no life existed in the great dark deep. And yet several other expeditions—including John Ross's to the North Pole in 1818—had plumbed depths of almost 6,500 feet and delivered proof of a rich and varied animal life below.

A man on a small windswept island off Vestlandet, the southwest coast of Norway, also proved that Forbes was full of shit. Michael Sars and his son Georg Ossian Sars were among the first in the world to prove with scientific certainty that the ocean depths are not lifeless underwater deserts. They are two of the most formidable scientists Norway has ever produced. Their efforts are no less impressive when you consider where they started. Michael Sars came from modest circumstances in the west coast town of Bergen. It was out of the question for him to set his sights on a career in the field that was his true passion, which was life in the sea.[13] Instead, he went to Oslo, became a theologian, and married Maren Welhaven, the sister of the famed Norwegian author Johann Sebastian Welhaven. In 1831, he was appointed pastor on the island of Kinn, which lies on the northwest coast, just outside Førdefjorden. There Sars devoted all his free time to studying marine life. By 1835 he made a breakthrough with his work *Beskrivelser og iagttagelser over nogle mærkelige eller nye i havet ved den bergenske kyst levende dyr* (Descriptions and Observations of Some Strange or New Animals Living in the Sea off the Coast of Bergen). In recognition of the rare talent possessed by Sars, the Norwegian parliament, the Stortinget, awarded him a stipend. This allowed him to travel throughout Europe and enabled him to make contact with the foremost naturalists at the universities in Paris, Bonn, Frankfurt, Leipzig, Dresden, Prague, and Copenhagen. In the early 1850s, using a rowboat and scraper, Sars studied the vast depths of the Mediterranean. He found life as far down as twenty-six hundred feet. That was as deep as he got.

One of the many people fascinated by Sars's discoveries was Peter Christen Asbjørnsen, who would later become world famous as the collector of Norwegian folk tales, along with Jørgen Moe. While Asbjørnsen sought out isolated mountain valleys in search of old tales, his thoughts must have been elsewhere at least part of the time because he wanted to become a marine biologist. And his role model was Michael Sars. In 1853 Asbjørnsen published a treatise entitled *Bidrag til Christianiafjordens litoralfauna* (Contribution to the Littoral Fauna of Christiania Fjord). It dealt with various life-forms in the intertidal zone of what is today called the Oslo Fjord. But what really fascinated Asbjørnsen was life in the ocean deep.

In the same year that his treatise was published, and with a state-sponsored stipend in his pocket, Asbjørnsen traveled to Vestlandet to study the deep fjords in that area. He first went to visit Sars, who was then the pastor in Manger, on the island of Radøy in Nordhordland. Asbjørnsen had worked to establish a special professorship, which was actually intended for Sars. After convincing the pastor to apply for the position, Asbjørnsen started on his own marine biology investigations. The results attracted the attention of zoologists.

Asbjørnsen managed to bring up an eleven-armed starfish from a depth of thirteen hundred feet in Hardangerfjorden, using a bottom scraper of his own construction. The coral-red starfish, "shimmering like mother-of-pearl," was a new discovery. As the person who had discovered the starfish, Asbjørnsen was entitled to give it a name. He called it *Brisinga endecacnemos,* after the Brísingamenet, the beautiful brooch that, according to Norse mythology, supposedly belonged to the goddess Freya. But it was the trickster and shape-shifter Loki who had brought it up from the bottom of the sea.

Asbjørnsen thought his jewel of a starfish was a unique species, but he bowed to the opinion of Michael Sars, who expressed doubts about this. Later it turned out that Asbjørnsen was correct, but he had been cheated out of the recognition for his discovery.[14]

In spite of his diligence, Asbjørnsen for the most part did not receive the stipends and positions he sought. His career as a marine biologist stagnated and finally came to a halt. He had to make new plans. Forests also held a strange attraction for him, so in 1856, he went to Germany to study at the Royal Saxon Academy of Forestry in Tharandt. He graduated with the highest marks in all subjects and became a progressive force within the administration of forests and wetlands in Norway.[15]

Yet sometimes people do receive the recognition they deserve. The great German evolutionary biologist Ernst Haeckel wrote of Michael Sars, "For all those who had the joy of knowing him personally, the liveliness of his spirit, the kindness of his disposition, the clarity of his mind, and the versatility of his knowledge will remain unforgettable."[16] The first Norwegian oceanographic ship was named after Sars. The newest ship used by today's Norwegian oceanographers—which is brimming with advanced technology and extremely quiet so that the engine noise won't disturb the acoustic instruments—was named after Sars's son, Georg Ossian.

He carried on the work of his father, who, with great tenacity and thoroughness, had paved the way in Norway for the field of marine research. In 1864, Georg Ossian Sars was the first Norwegian to receive a government-sponsored salary, with the title of "marine researcher." The same year he traveled to Lofoten, or more specifically to Skrova, which served as his base as he brought up great numbers of samples from the deep waters of Vestfjorden.

In 1868, G. O. Sars published his results, which attracted the attention of the international scientific community.[17] Of special interest was what became known as Sars's sea lily (*Rhizocrinus lofotensis* Sars). The Lofoten sea lily was described by Sars as a "living fossil" at a time when scientists were combing the earth in search of such things in order to substantiate the theory of evolution and to date the age of both the earth and life itself.

Even so, it took a long time before the discoveries of complex life-forms in the ocean depths became generally accepted. When a telegraph cable was laid along the floor of the Atlantic Ocean in 1858, one engineer who worked on the project claimed that both starfish and globigerinae (a type of plankton that exists in great quantities on the seafloor) were attached to the plumb line when it was pulled up from depths where nothing could live, according to what was known at the time. Most scientists took a skeptical view of such claims. Some maintained that the animals must have become attached on the way up, even though many were clearly bottom dwellers. But the fuse had been lit, and some of the discoveries were impossible to ignore.

Both Asbjørnsen's starfish and Sars's sea lily were cited when the leading Scottish zoologist Charles Wyville Thomson applied to the Royal Society in London for funding for his expedition on board the *Lightning* in 1868. The purpose was to explore the deep sea areas off of Scotland. The expedition confirmed and expanded upon the discoveries made by the Norwegians. Extremely interesting life-forms were found as far down as four thousand feet.

In 1872, C. W. Thomson was an obvious choice to participate when the British prepared for the first major, modern sea expedition. With a crew of 270 (including officers and scientists), the HMS *Challenger* sailed the oceans of the world for four consecutive years, the whole time plumbing the depths, mapping the currents, and measuring the temperatures. Open waters were trawled at varying depths, and samples were brought up using the same method Michael Sars had developed.

The results of the *Challenger* expedition formed the basis for modern oceanography. No one could continue to insist that the vast ocean depths were lifeless zones now that even the most renowned (and British) scientists claimed the opposite. Exactly what existed on the seafloor was hotly discussed, also in the media and by means of articles popularizing the topic. For instance, *Skildringer af Naturvidenskaberne for alle* (Depictions of the Natural Sciences for Everyone)[18] from 1882 contains translated articles by leading European scientists and experts. The ocean depths, in particular, are given great attention. The Englishman Philip Herbert Carpenter, who was an expert on sea lilies and who also took part in the *Challenger* expedition, begins like this: "For most of us, the deep sea floor is a completely unknown territory because its location makes it impossible for us to ever undertake a direct and personal exploration of its marvels." Carpenter was a talented but tormented man. Chronic insomnia drove him insane, and in 1891 he used chloroform to take his own life. But Carpenter managed better than most others before him to see for himself the underwater landscapes. "Our investigations have taught us that in many ways the sea floor, with its enormous expanses, offers a resemblance to the surface of the earth. Similarly, it has its mountains, valleys, and vast undulating plains. Its components vary greatly in different places; it has its deserts and its fertile regions, its forests and its cliffs, and like the earth's surface, it is inhabited by various animals and plants in the different areas and climates."[19]

For close to a hundred years after Carpenter wrote this, the prevailing opinion was that there was very little variety of life on the seafloor, that it consisted mainly of sea cucumbers, worms, and smaller animals. Even today only a few underwater submersibles can reach the deepest depths. With every new expedition they discover not only new species but also life-forms previously unknown. The same thing happens each time scientists lower nets or scrape the bottom at great depths that haven't been explored. In fact, the majority of the species they bring up have never been described before.

The deep, which until recently was thought to be dead, is actually teeming with life. It's pitch-dark, but most of the species produce their own light, in every imaginable color and variation, in order to lure or entice others. The deep is constantly flashing and glowing. More species live down there in the dark than on land, making their language of light the most widespread means of communication on earth. Thousands of feet below the surface, the most absurd creatures exist. For example, the humpback anglerfish, or blackdevil, which has an illicium, or "fishing lure," arcing from the top of its head (or from the lower jaw), with a light at the end that dangles in front of its eyes. The fish floats motionless in the water column with its huge, open jaws, from which protrude long, sharp teeth. Its body is covered with hundreds of long antennae, which allow the fish to sense the slightest movement in the water. If anything comes close, the fish lunges for it.

Many species are nearly as transparent as glass. Only the small digestive organ gives them away if caught in light. If they're aware of danger, some are capable of pumping great quantities of seawater into their body in order to make themselves even more transparent. Some life-forms are round and lack a head. Others, belonging to the Siphonophorae, look like strings or ribbons of pulsing plasma dancing around, in a seemingly graceful and coordinated movement. A type of jellyfish colony, *Praya dubia,* can reach 130 feet in length and has three hundred stomachs. The Dana octopus squid (*Taningia danae*) has big light organs on all eight arms, and when it's hunting, which it does in packs, all of the lights can flash at the same time. The prey must think it's being attacked by huge Christmas ornaments. Another deep-sea squid (*Heteroteuthis dispar*), called the "fire shooter" by those who have nicknames for cephalopods, is able to shoot out clouds of light to confuse predators.[20] If the jellyfish *Atolla wyvillei* is attacked, it flashes thousands of blue lights, like an emergency vehicle. The light show can blind or confuse the attacker or even attract bigger predators, which swallow up bewildered spectators and eliminate the danger to the jellyfish.

Most bioluminescence produced by various deep-sea species is blue, because blue is the color that penetrates farthest down in the water. That's why the ocean looks blue. Blue light is the only type of light that most species in the deep are capable of seeing. The smalltooth dragonfish (*Pachystomias microdon*) has developed red lights in addition to blue. Using the red light, the fish can approach other animals that have no clue a spotlight is being directed at them. Another type of dragonfish is called the *Malacosteus niger* in Latin, but let's call it the "wobblemouth." Its lower jaw is as elastic as a slingshot, enabling it to fire its mouth at prey faster than the eye can see.

Many species use light to find a mate. It's not the safest thing to do, because when they send out their signals, they also draw the attention of predators. Some have developed cunning mechanisms that resemble the mating signals of other species, so as to lure them close and devour them.

In the ocean, enemies can come from every direction, and at any time. That's why many animals that live hundreds of feet down in the water column have camouflage lights on their belly, enabling them to blend in with the water, whether seen from above or below. This is a wily defense mechanism, but it's also a contrivance that can give them away. The eyes of some species can differentiate the artificial, bacteria-formed light from the prey swimming above, so its silhouette is no longer invisible.

A sea cucumber that can live as deep as sixteen thousand feet sheds its own skin if attacked. The skin is sticky like double-sided tape and will keep the attacker occupied while the sea cucumber makes its escape. Others use poison or barbs. No one ever claimed that life in the deep is simple or pleasant.

More people have gone up into space than into the vast ocean depths. We are far more familiar with the surface of the moon, and even with the dried-up seas on Mars. But if we could swim around down there in the cold and dark, it would be like floating in outer space, surrounded by twinkling stars. Brilliantly colored fish using arms to walk on the seafloor. Yeti crabs dressed in white fur. Hairy anglers (*Caulophryne polynema*) with fishing poles on their heads, swaying back and forth like the pendulum on a metronome, with a seductive light on the end. No fish shines brighter than the illuminated netdevil (*Linophryne arborifera*), which has a long antenna sticking up from its snout and a shrublike appendage called a barbel hanging from its lower jaw. In this case we're talking about the female, because the male is merely a small parasite, which early in life attaches itself to the female's belly. That's how it spends the rest of its life, receiving nutrients from the female's blood and in return regularly donating sperm.

The giant squid (*Architeuthis*) glides horizontally through the water at great speed, possibly only a few yards above the seafloor, with its arms gathered behind in an aquadynamic point and eyes as big as plates that never blink. It's equipped with water-jet propulsion and camouflage systems that the U.S. Navy would love to imitate.

Organic material is constantly raining, or rather snowing, down through the entire water column. A dizzying number of specialized creatures make use of everything that drifts down to them.[21] Over the last few years so many new species have been discovered in the deep just by taking random samples that some people believe this ecosystem alone could contain several million species. It's true that the majority of life in the sea keeps to the uppermost layer of water. But for all we know there are more species in the depths, and almost all life down there possesses astonishing characteristics, as if belonging to a different planet or created in a distant past when other rules applied and any fantasy could be realized. Down in the deep, life is like a dream from which it takes a long time to awaken.

7

At the halfway point across Vestfjorden I ask Hugo to stop so I can take off my thermal suit. I've never before experienced heat as a problem in these waters. The Lofoten Wall is getting closer, but the haze makes it blurry, as if the mountains have grown soft and are starting to melt.

As soon as Hugo gets the boat going again, I catch sight of a column of water shooting straight up out of the sea, many miles ahead, slightly to starboard. I turn around and signal to Hugo, who nods and ramps up to full speed. We quickly approach what now looks like a shallow islet polished smooth and gleaming in the sun. But we're in open ocean, where there are no islets. And besides, this one is moving. We've already seen several porpoises, but this is clearly something else entirely. Hugo starts speculating out loud.

"Well, it's not minke whales, at any rate. Could it be a pod of pilot whales?"

We're several hundred yards away when Hugo realizes that this can't be right either. What we see in front of us has no dorsal fin, like the pilot whale has. And it's not a pod at all. It's one huge animal. For a split second I wonder if it might be a submarine. Hugo's body is tensed, his gaze fixed, his mouth open as he frantically pages through his internal catalogue of various types of whales. We're only a couple of hundred yards away when he exclaims:

"It's a *sperm whale!*"

In front of us is one of the largest of all toothed whales. As we approach, it starts to arch its back. When we're a hundred feet away, it blows one last time and then lowers its head into the water. The flukes and hind part of the body stick vertically up from the surface, iconic as a rock carving, before the sea closes around them. The whale is gone, as if someone had pulled a string, drawing it down into the abyss.

Hugo switches off the motor. He has spent nearly fifty years by the sea, and so much time in Vestfjorden that he could almost be considered part of the fauna. During that time he has seen just about everything. Groups of pilot whales are practically routine, not to mention minke whales, dolphins, and porpoises. But not once in all those years has he ever seen a sperm whale.

Now it's just a matter of waiting. Even though a sperm whale can hold its breath for ninety minutes—longer than any other creature with lungs—eventually it will have to resurface.

The sperm whale (*Physeter macrocephalus*) is not only the largest carnivorous animal in the world. It's the largest carnivorous animal that has *ever existed on earth*. Forget the *Tyrannosaurus rex,* the megalodon shark, and the kronosaur. The sperm whale is both heavier and longer. Almost nothing that has ever lived, or is alive today, including the other big whales, can compare.

The animal we saw was a solitary male about sixty-five feet long and weighing more than fifty-five tons. Males and females are not alike. The females weigh only a third of what the males weigh, and they live in groups that take care of the calves, sharing the responsibility for another female's calf when one of them dives to search for food. Young males swim around in groups. Puberty is over when they reach the age of thirty. By then the male sperm whale has had enough companionship to last the rest of its life. From that moment on, the male becomes a solitary hunter in the oceans of the world. The whale we encountered could have swum all the way from the Antarctic Ocean. If it meets a group of females, it might mate, but as soon as that's over, it will take off again. The males can be aggressive when they run into other sperm whales. Maybe it's sexual frustration that makes the males fight, even in a sober state. Hugo says a horny sperm whale can get as crazy as an elephant in heat.

While waiting, I ponder what the sperm whale that disappeared into the deep might be up to. It could be hunting octopus or a giant squid, which can weigh up to a thousand pounds. On its way down, the whale might grab the octopus in its teeth and crush it against the seafloor. If the whale doesn't find anything as it descends, it has another chance to nab food while going back up. At the bottom, the whale starts swimming upside down, scanning the water above for shapes silhouetted against the faint light from the surface. The sperm whale uses the sonar system at the front of its head to localize schools of fish or squid. If the whale spots something interesting, it speeds up, swallowing the prey in a mouth big enough to hold crosswise the boat in which Hugo and I are sitting.

Sperm whales that have washed ashore have been found with deep sucker marks, sometimes eight inches in diameter, on their bodies. No humans have ever witnessed a battle between a giant squid and a sperm whale, but if the opportunity arose, the tickets would sell out instantly. The giant squid, which was long regarded as a fictitious monster, not only has eight arms, each of which can reach a maximum length of twenty-five feet; it also has a repulsive bony beak that can crush just about anything. According to Jules Verne, the arms of this colossal aberration are like a Fury's strands of hair. It should be easy to make eye contact with the giant squid, because it has huge round eyes, with no eyelids, so it never blinks.

On the front of its head the sperm whale has the largest sound-producing organ in the animal world. It can weigh eleven tons all on its own. The clicks it makes have been measured to 230 decibels, a sound level comparable to that of a rifle shot fired four inches away from your ear. The males emit loud bangs like this, while the females speak more rapidly, almost like a type of Morse code.

As an evolutionary heavyweight, the sperm whale ought to be swimming around with an enormous silver belt around its stomach. But even the sperm whale has its enemies. It gives birth to few offspring, fewer than any other type of whale, and many years are required to teach, feed, and protect the calves. Both calves and injured adults can be attacked by groups of orcas or pilot whales. In those situations the sperm whales take up a so-called marguerite formation, in which all the bigger animals make a circle around the calves. This way the sperm whales can turn in any direction and use either their tail or teeth as weapons against their attackers. This can prevent the much faster and more agile orcas from isolating the calves, which would spell the end for them.[22]

The sperm whale can dive to a depth of close to ten thousand feet, which is a record for mammals.[23] At such depths, their lungs are nearly compressed flat. Inside its head the whale has a large chamber where the pressure is equalized as the spermaceti oil cools down, solidifying and attaining greater density during the dive for the bottom. The oil warms up to a liquid state closer to the surface, enabling the whale to float. Until about a hundred years ago when synthetic replacements were invented, spermaceti oil was the most valuable of all oils. It is pure, transparent, and pleasant smelling. A big whale could contain up to five hundred gallons of spermaceti oil in its head. The finest candles, soaps, and cosmetics were made from this pale pink, waxy, spermlike liquid. And spermaceti oil was also used to lubricate the most expensive precision instruments.

Many other parts of the animal were also of great value. A single whale could produce several dozen tons of blubber and meat, and the huge teeth were as precious as elephant ivory. It's said that whale hunters even made themselves raincoats from the skin of the enormous penis. That's not the only way in which the sperm whale is endowed. It also possesses the largest brain of any animal on earth—ever. The brain weighs six times more than a human's. The penis weighs several hundred times more.

To top it all off, the sperm whale secretes the substance called ambergris in its digestive tract. Ambergris was the most valuable part of the whole whale, since it was used in perfumes. Many people ascribed all sorts of fantastical qualities to the substance. In the old days, when ambergris was found floating at sea or washed ashore at low tide, people thought it was something a sea monster had spit out. Hugo has found ambergris, which they used to call whale amber, on the intertidal zone. He describes it as waxy gray lumps with a characteristic, slightly sweet smell.

The sperm whale was in high demand and was hunted so vigorously that it came close to extinction. Off of Andenes, an important fishing station at the northern tip of the Norwegian island of Andøya, this type of whale was systematically hunted well into the 1970s. Before whale hunters had grenade harpoons, they used big harpoons that went right through the whale, before catching on barbs. But the hunters still lost a lot of animals. If the whale's vital organs weren't damaged, it could swim around for years with the harpoon buried in its body.

All is quiet around me and Hugo, except for a soft, musical sound from invisible currents lapping against the boat. Elsewhere around us the water licks at the underside of its own surface, which gleams above the hollows and high shoals.[24] The sea is a cohesive sheet of light, so bright that it seems to illuminate itself. To the west the ocean swells in a convex arc, like an overstuffed dumpling. We're looking at the curve of the earth. There's still no sign of the sperm whale, and if it were an ordinary day, there would be little chance of seeing it again. But this is no ordinary day. The sea is so calm and the weather so clear that we'd probably be able to spot the colossus of a sperm whale at a distance of five miles.

Hugo tells me about a local incident from the previous century, when a sperm whale attacked a small boat carrying a large family to church. They were going from Lottavika to Leines when the sperm whale shattered their boat. Only a sixteen-year-old girl survived; the rest of the family drowned. Apparently air pockets in the girl's dress kept her afloat.

The core of the story is true, but local historians think the sperm whale happened to collide with the boat while it was grazing for herring.

On the other hand, it was no accident in 1820 when the whaling ship *Essex* of Nantucket was attacked by a sperm whale in the southern Pacific Ocean. The whalers on board the ship, which was eighty-seven feet long, estimated the whale to be eighty-five feet in length. They'd never seen such a large animal. For a long time the whale swam peacefully a good distance ahead of the *Essex*, as if keeping an eye on the ship. Suddenly the whale turned and headed full speed toward the ship, ramming it with tremendous force and smashing a huge hole in the bow. The crew on deck were knocked off their feet. A moment later the whale again attacked, shattering the other side of the bow into kindling. The whale kept at it until the 262-ton ship sank. First mate Owen Chase and more than half of the crew survived. Chase gives a sobering account of what happened in the book *Narrative of the Most Extraordinary and Distressing Shipwreck of the Whale-Ship Essex* (1821).

This isn't the only documented incident in which great ships have been sunk by sperm whales. But the story of the *Essex* is the most famous because it inspired Herman Melville to write his book about the white sperm whale Moby-Dick. The book is filled with chronicle-like chapters on whaling and the anatomy and behavior of whales ("The Sperm Whale's Head," "Measurement of the Whale's Skeleton," "Does the Whale's Magnitude Diminish?," and so on). According to Ishmael, the narrator, the white whale becomes for Captain Ahab the manifestation of all the evil forces that certain "deep" people feel are eating away at them:

> That intangible malignity which has been from the beginning; to whose dominion even the modern Christians ascribe one-half of the worlds; which the ancient Ophites of the east reverenced in their statue devil;—Ahab did not fall down and worship it like them; but deliriously transferring its idea to the abhorred white whale, he pitted himself, all mutilated, against it. All that most maddens and torments; all that stirs up the lees of things; all truth with malice in it; all that cracks the sinews and cakes the brain; all the subtle demonisms of life and thought; all evil, to crazy Ahab, were visibly personified, and made practically assailable in Moby Dick.[25]

The captain was insane, and his madness was contagious. The whale became the enemy of the entire crew, and almost to the same extent. Although Ishmael doesn't realize it—because here it's the author Melville speaking directly to his readers—Moby-Dick is the crew's "unconscious understandings," the "gliding great demon of the seas of life":

> The subterranean miner that works in us all, how can one tell whither leads his shaft by the ever shifting, muffled sound of his pick? Who does not feel the irresistible arm drag?[26]

The whole crew follows Ahab's lead, because the same thing exists inside all of them: an inherited and instinctive murderous force that is destructive of both the world and everyone around them. It's also self-destructive. Moby-Dick is the mammal threatened with extinction that was slaughtered by the tens of thousands in Melville's day, and it's also the darkest of forces in human nature. Like the desire for revenge, or the monomaniacal search for truth and the control of "innocent" nature. Ahab is the one hunting the whale, not vice versa. In the end he is dragged down to the bottom with the line from his own harpoon wrapped around his neck. That is how he is finally united with the Great White Whale.

More than two hundred million whales, of varying species, were caught globally from the 1870s to the 1970s. Over the course of a few decades, the local whale populations, which had numbered in the tens of thousands, ended up reduced to a few terrified animals.[27] Norwegian companies based in Larvik, Tønsberg, and Sandefjord carried out commercial whaling for more than fifty years in the Antarctic Ocean as well as off the coasts of Australia, Africa, Brazil, and Japan. Huge factory ships were built at Norwegian dockyards, and hyperefficient processing furnaces (historically called tryworks), were transported to South Georgia Island in the southern Atlantic Ocean and Deception Island in Antarctica. In 1920, on Deception Island alone, there were 350 pressure cookers, each capable of holding twenty-six hundred gallons of oil. Before the blue whale came close to extinction, the whaling ships caught thousands of them every season, as well as a number of other species. Live fetuses were cut from the wombs of pregnant female blue whales and incinerated. The days were not calculated in hours but rather in the number of whales caught and barrels of oil produced. The smoke and steam from the enormous, roaring furnaces hovered like a thick blanket over the whaling stations. A single blue whale can have two thousand gallons of blood in its body, and the men who did the flensing waded nonstop in blubber, blood, and meat during the four-month season.

The stench of death and putrefaction was indescribable. The furnaces and factory ships were often unable to keep up, which meant that whales were left lying on the foreshore until their flesh turned rancid and the gases inflated their bodies like zeppelins. When these carcasses were pierced, or if they exploded on their own, the stench was enough to make people pass out. The surrounding shores were like gigantic whale graveyards, where thousands of carcasses, skeletons, and bones lay rotting. Some people claimed they could never get rid of the smell; it lingered in their nostrils even decades later.[28]

All whales are able to communicate with one another over great distances, but increased ship traffic is making that more and more difficult. Yet this problem is surmountable compared to what the "world's loneliest whale" has to endure. Fin whales normally communicate at a frequency of twenty hertz, and they hear only sounds close to that frequency. But several years ago whale researchers were astonished to discover a fin whale with a special handicap: it sings at a frequency of approximately fifty-two hertz. This means that none of the other fin whales can hear it, and the whale is cut off from all social interaction with its fellow fin whales. Maybe the others think it's mute or a different species or an asocial eccentric. The "world's loneliest whale" keeps to itself. It doesn't even follow the migration routes taken by the others through the oceans of the world.[29]

As a boy, Hugo often went out on the *Kvitberg II,* a boat equipped for all types of fishing. In the designated season, it also went after whales. Once, from the dock, Hugo watched the crew cut open a pilot whale after the ship returned from a hunt. They hit a blood vessel, and as he remembers the incident, he saw columns of blood spraying across the deck. But he has begun to doubt this memory, because the whales would have been cut up into huge chunks weighing sixty-five pounds each by the time the *Kvitberg* returned from the Barents Sea. Could he have seen a whale that had been caught in Vestfjorden? At any rate, the blood vessels on the inside were as thick as cables and very visible as the heart was sliced in half. Men stood ready on the dock in Helnessund with meat hooks, which they stuck into the pieces and then dragged them along the wharf into the refrigerated warehouse.

—

Where has our sperm whale gone? The waters all around us are suddenly swarming with herring. The surface is so smooth that from far away we can see the massive schools of fish. If we'd had a seine, which of course would have required a much bigger boat, we could have easily pulled on board many tons of herring. Seabirds hover over the fish, eating so much they can hardly take flight. We see northern fulmars (which belong to the Procellariidae family), great cormorants, common eider, and ordinary seagulls. Even an arctic tern, flying low, passes our boat. Every year they fly from the South Pole to the North Pole and back.

The gentle whispering of the sea, the dry warmth of the sun, the air so clear—such peace. These are the kinds of days that a person collects, to be recalled years later. Only one thing spoils the idyllic scene. The Scottish Highland bull. The smell is seeping through the sacks. Apparently it wants all of Vestfjorden for itself. Several seabirds turn away in midair as they get close to our boat. Others make odd maneuvers, as if for a brief moment they actually faint. It's been almost forty-five minutes. Could the sperm whale already have surfaced so far away that we didn't see it, only to dive down again?

Hugo and I are in the middle of discussing the origins of the Norwegian phrase "drunk as an auk" when we hear a booming sound in the distance. We sit perfectly still and listen. There it is again.

"It sounds like a rock slide. They must be doing some blasting on shore," says Hugo, turning to look toward Kabelvåg.

Again a thunderous sound rumbles across the surface of the water. It reminds me of the deepest tones from a church organ, but wetter and with a trace of heavy gurgling. This is not the sound of blasting operations on shore. It's the whale pumping air in and out of its giant lungs.

"There!" says Hugo, pointing north with one hand as he turns the ignition key with the other. A spout of water breaks the surface far away, and Hugo gives the boat full throttle. Several minutes later we're close to the whale. It's hardly moving at all, just breathing. Each time it exhales, it roars, and a spout, like water from a fire hydrant, leaps from its blowhole, which is at the front of its head, on the left. We can hear the air sucked into its lungs, like the whoosh from the open window of a speeding car. In between, there are loud booms. It's what Rimbaud called the "moaning of the Behemoths in heat."

The whale rolls back and forth a bit, showing us its strange knotty surface. It's as big as a bus. The part of the whale that's visible is almost double the size of our boat. Underwater we can glimpse the top of the whale's head, which has the same shape as the Kola Peninsula. Its eyes are so far down in the water that we can't see them, but they undoubtedly see us.

After traveling in Africa, India, and Indonesia, I've become rather blasé when it comes to experiencing nature and wildlife. But right now I'm simply sitting here and staring, dumbstruck by the size and power of this creature. Then I finally come to my senses and grab my camera.

Hugo steers the boat even closer, and I start to feel anxious. What if the whale gets annoyed and decides to give us a swat with its tail? We'd be launched high into the air, with the heavy outboard motor and propeller still going. It's a long way to shore, but Hugo thinks we're safe as long as we stay alongside the front of the whale.

Almost everyone has heard the story of Jonah and the whale.

George Orwell, in his essay entitled "Inside the Whale," has also written about this same experience, although in a figurative sense:

> The historical Jonah, if he can be so called, was glad enough to escape, but in imagination, in day-dream, countless people have envied him. It is, of course, quite obvious why. The whale's belly is simply a womb big enough for an adult. There you are, in the dark, cushioned space that exactly fits you, with yards of blubber between you and reality, able to keep up an attitude of the completest indifference, no matter *what* happens. A storm that would sink all the battleships in the world would hardly reach you as an echo. Even the whale's own movements would probably be imperceptible to you. He might be wallowing among the surface waves or shooting down into the blackness of the middle seas (a mile deep, according to Herman Melville), but you would never notice the difference. Short of being dead, it is the final, unsurpassable stage of irresponsibility.[30]

After what must be about three minutes, although it feels like fifteen, the sperm whale gets ready to dive. It arches the enormous front part of its body in several preparatory maneuvers. We're ten or twelve feet away when the whale points its snout downward and its body slowly follows until its crescent-shaped fluke is sticking out of the water in front of us. It vanishes without a sound.

Then something odd happens. In front of our boat, maybe sixty feet from the spot where the sperm whale dived, the water starts quivering with tiny ripples and currents, as if it's a high-voltage electrical field. The whale is heading our way. I look at Hugo, no doubt with panic in my eyes. He has noticed the same thing. He sets his hand on the throttle lever and gently veers away from the mass of pure power that is coming toward us.

Suddenly everything goes calm, and the entire sea is once again as shiny and smooth as blue chrome. The sperm whale is on its way down to the deep.

Hunting for a Greenland shark? After our encounter with the sperm whale, it now feels like we're setting off on a perfectly ordinary fishing trip.

8

Now our search for the Greenland shark begins in earnest. Studying our sea charts, we triangulate our position using landmarks on shore: the Skrova lighthouse; an old, cone-shaped stone marker on the outermost islet; and Steigberget, at the top of the glacier Helldalsisen on the other side of the fjord. We've reached the approximate spot where we planned to try our luck. I stick a hole in the trash bags that are filled with intestines, kidneys, liver, gristle, bone fragments and joints, fat, sinews, fly larva and maggots. I throw up nonstop. As I mentioned, Hugo is incapable of vomiting, but he looks like he wished he could, positioned as he is at the far end of the boat, leaning over the side. Then I dump four of the five sacks over the gunwale. There are rocks in the bottom of the bags, and they sink toward the seafloor. The fifth sack contains some meaty tidbits that we're going to use as bait on the hook.

The sea is at least a thousand feet deep in this spot. I've read in books of local history that the fishermen used to wait twenty-four hours before coming back to try and get the Greenland shark to bite. We're going to do the same thing, even though it really shouldn't be necessary. If there's a Greenland shark within a radius of five miles, it's only a matter of time before the shark catches a whiff of the goodies way down there in the deep. Like other sharks, the Greenland shark can smell "in stereo" and pinpoint the source with great precision. And even though there are no waves in the sea, the currents off Skrova are always strong. In this particular spot, the current will spread the smell of the offal as if it were blowing in the wind. That's our theory at any rate. Tomorrow we'll see if there's any truth to it.

Fishing for Greenland shark resumed during World War I. Poor folks ate the meat, and the livers could be used for lamp oil, medicinal oils, and many other things. Hugo's great-grandfather Norman Johan and his sons Svein, Hagbart, and Sverre were among the first in the district to process oil from Greenland sharks. In other words, it's in Hugo's blood. If anyone should take up the tradition fifty years after the fishing for Greenland sharks ended, he's the most likely candidate.

We steer the RIB close to the Skrova lighthouse, which stands on a small rock of an island. We roar through more schools of herring. They leap around the boat in glints of shiny silver. Here the sea is never totally still, but on this day it's as close to calm as it gets. Near to shore, I can see nearly imperceptible waves race toward the bare rocks, soundlessly, and without breaking. The water moves lazily, as viscous as floating aspic.

Just off the islet that the locals call Kvalhøgda (Whale Heights), we let out a normal fishing line to catch something for dinner. I can actually feel fish stop the swaying descent of the lead weights toward the bottom. Herring at the very top, grazing on zooplankton. Under the herring swim pollock, also going after zooplankton. Below the herring, zooplankton, and pollock there are bigger fish. A halibut snaps at one of the pollock we have on the hook, ripping off its skin, though, unfortunately, without getting snagged itself.

After entering the small sound between the islets of Saltværøya and Skarvsundøya, we head for Skrova, which isn't really a single island but a small cluster of islands and islets. For a hundred years the fishing village of Skrova has been a hub for the area's fishing and whaling. The reason for this is both geographic and topographic, since the island group is situated out in the ocean, almost in the middle of the fishing banks and whale hunting grounds in Vestfjorden. And Skrova has a good, safe harbor.

Today well over two hundred people live on Skrova. The fish landing stations have been shut down, except in the Lofoten season, but there is a processing plant for farmed salmon on the island. And almost all minke whales caught in Vestfjorden in the springtime are still taken to Skrova, to the modern plant operated by Ellingsen Seafood.

Skrova has a natural harbor, with a snug entrance into a bay that's the perfect size in length and width. In Heimskrova, the buildings are set close together, giving the community a more intimate and country-village feeling than is usual this far north. Traditionally, wherever there was more space—for instance along a fjord—the northerners in Norway would build their homes far apart, especially since each dwelling had to have space for a small farm, with fields, cowsheds, and maybe a pasture for grazing, as well as a boat berth on the foreshore. On Skrova there is very little pastureland, and most of the buildings stand close together in the fishing village and on the surrounding islets. In this raw landscape, there's comfort in company.

The island is always bathed in light from the sea, and when we come racing into the bay in our RIB, Aasjord Station is the first thing we see. Standing on posts on the small islet of Risholmen, and with the sea on three sides, it's naturally what catches the eye of everyone who arrives. At this time of year, the sun shines on Aasjord Station 24/7. It feels as if the fishing station turns with the sun.

The last time I was here, the place looked as though it were about to fall into the sea. The dock and posts were rotting. Decades of neglect had brought the structure close to collapse.

Now it smells of fresh lumber and linseed oil. The whole wharf is brand-new. The posts holding up both the wharf and the buildings are made of aspen, which won't rot in seawater. The buildings have been given new board siding, painted white, which makes them visible from many miles away. In the background the black peaks of Lillemolla stick up from the sea. It's no wonder that Christian Krohg hesitated to set up his easel under such conditions. When another Norwegian painter, Lars Hertervig (1830–1902), was asked by his doctor what had caused him to lose his mind, Hertervig said that he had "stared too much at landscapes in strong sunlight" and he "lacked good paints" to depict the landscapes in an accurate manner.[31]

—

Hugo and Mette live on the second floor of one of the buildings, in a small, two-room apartment that was created in the 1970s for workers at the fishing station. Except for a few such units serving as living quarters, most of the building consists of large open spaces. There are tons of fishing lines, nets, seines, and all sorts of other things needed to operate large fishing boats, a fish landing center, and a cod-liver-oil mill. Along two sides of both buildings, dormers jut out from the attic, with big double doors so that everything can be hoisted up or lowered down directly from or into the boats.

The roof, wharf, and outer walls have now been secured. In a few years the interior will be completely refurbished. The plan is to turn the station into a restaurant, overnight accommodations, and a retreat for artists. Hugo also wants to start up a small private fish landing center, to show visitors how the raw material was handled in the old days. It's a bold and expensive venture that will require support from many sources, including the goodwill of the banks, if it has any chance of succeeding. Mette and Hugo have already mortgaged their house in Steigen. They're facing many years of hard work ahead, and there's always a risk of failure.

Skrova isn't just close to the sea. It's *in* the sea. Even Aasjord Station stands on posts and belongs to both the land and the water. On shore the only access to the station is via the neighbor's dock. During spring tides, with low pressure systems and wind from the west, the sea can rise so high that it looks like the whole station is floating.

"The house is like a conch shell, surging with the roar of the sea. Ceaselessly the ocean treads toward land, today as it did yesterday."[32]

9

In the evening Hugo, Mette, and I decide to pay a visit to Arvid Olsen, the oldest fisherman in Skrova. Like everyone else in Skrova, Olsen lives in a detached house. His is on the outskirts of the village, and he's been living there since the 1950s. It's a small, cozy-looking place, with a nice little garden sheltered by a boulder. And in Skrova wherever there is shelter—at the foot of mountains and cliffs—you'll find a surprising variety of trees, decorative shrubs, and plants. Many have been brought here from the south. Others, like the maple trees and Persian hogweed, arrived from the east via the Pomor trade—carried out between the Pomors of northwest Russia and the Norwegians along their northern coast as far south as Bodø. The plants were imported by the village's wealthy skippers and fish buyers. In the 1930s, a sailor brought a lily all the way from Australia, and the lilies still grow in some people's gardens. You wouldn't think such plants could survive this far north, but out here in the sea, freezing temperatures seldom last long.

One strange thing I'd noticed when passing Olsen's house was that the curtains always seemed to be drawn, which is unusual for a place like Skrova. But I hadn't asked Hugo about it. No one answers when we knock on the door. Finally we step inside and try knocking on the kitchen door. Olsen comes out of the living room and says he heard someone knocking, but only folks from the south ever do that. And since he knows we're not from the south, he figured it couldn't be us.

On the table we see a cake that his son and daughter-in-law brought over earlier in the evening because today is Olsen's birthday. He's getting close to ninety, but that doesn't seem possible when I look at him. As if to underscore his youthfulness, Olsen jabs out his hand in an attempt to catch a fly in midair.

Olsen was a fisherman from his teen years until he turned sixty-five. He caught cod, rosefish, pollock, and halibut, using a hand line, a long fishing line with dozens of hooks, and a net. But the most fun he ever had fishing was going after tuna. A big Atlantic bluefin tuna could bring thirty Norwegian kroner for everyone on board the boat. In comparison, he was sometimes paid twenty-seven øre for two pounds of the best Norwegian arctic cod. Olsen says that they ate only the meat located along the jaws of bluefin tunas.

Illness forced Olsen to stop fishing twenty years ago. After a heart operation, he developed a rare allergy to sunlight. That explains why his windows are covered with a black film that keeps out ultraviolet light. In the summer he can hardly leave the house without burning his sensitive skin.

We've come to hear about the Greenland shark. For Olsen, the shark has been mostly a nuisance. It often bit big pieces out of the halibut that landed in the net or took the bait.

"It eats everything it comes across. If you're fishing for Greenland shark, you have to pump the carcass full of air after you cut out the liver. If the carcass sinks, other Greenland sharks will attack the cadaver on the seafloor. They'll stuff themselves so full they won't be interested in the bait on the hook."

I accept his advice with a nod, without telling him that we'll be satisfied with just one Greenland shark.

"How much line do you have?"

"Thirteen hundred feet."

"Chain?"

"Twenty feet, attached to the end of the line."

"What are you using for bait?"

"A rotting Scottish Highland bull."

Olsen nods approvingly.

The way he talks reminds me of some of my elderly male relatives from Vesterålen who I used to meet as a child. Many of the terms he uses are special expressions used only by fishermen to describe the sea. For example, the word *høgginga* refers to when the currents begin to slow seventy-two hours after a full moon or a new moon. That's when the weather and winds often change. The word *skytinga* refers to when the currents start to increase after *småsjøtt,* meaning a low tidal range. In both instances, it's important to be out at sea when that happens, because that's when the fishing is good.

Over the next few days I make feeble attempts to master some of the old words. But they don't sound right in my mouth; it's like they don't belong to me. Nor do I fully understand all the nuances. So I decide to quit before I start getting on Hugo's nerves.

On our way home, Hugo and Mette tell me that the people living on Skrova have developed special ways to get extra protein. They eat the pickled, canned thigh of the great cormorant. And if they catch sea otters in their fish pots or nets, they cut fillets from them. This is not some long-dead custom. Mette heard about it from one of the kids at the grade school. In surprise, she asked, "Do you guys eat otter?" Four of the children nodded eagerly and told her it tastes delicious.

Back at Aasjord Station, Hugo gets out a small box. It contains photographs taken right after World War II by his uncle, Sigmund Aasjord, who'd been an amateur photographer ever since he was a boy. Hugo found the box in the warehouse of the old family fishing station in Helnessund. Many of the pictures were taken while fishing for Atlantic bluefin tuna (*Thunnus thynnus*). For several years after the war, there were huge quantities of this fish in Vestfjorden. The photos show seines filled with tuna. They can get to be eleven feet long and weigh as much as 650 pounds when gutted. As Arvid Olsen had told us, the prices paid for these fish in the Italian and Japanese markets were sky-high compared to what was paid for other types of fish caught in Norwegian waters. But this was also an entirely different sort of fish than Norwegian fishermen were used to. The tuna would die if they couldn't move in the seine. Then fifty or a hundred tons of dead weight would sink right to the bottom, causing a tremendous loss of revenue.

The bluefin tuna is one of the sea's most marvelous fish. Its whole body is like a single solid, powerful muscle, and the sleek, sickle-shaped tail can propel the fish at speeds reaching almost thirty-five miles per hour. Only a few other species, such as the swordfish, sailfish, orca, dolphin, and some types of shark, are faster. Most fish are cold-blooded, which means their body temperature changes according to the temperature of the ocean. But, like human beings, the tuna is warm-blooded and has a constant body temperature.

The tuna is capable of swimming from tropical waters to the Arctic and back—though there's a good chance it will be caught and killed along the way. Everything from helicopters to floating surveillance buoys and sensors are used to catch tuna. Fishing boats drift in the oceans with lines that are thirty to fifty miles long and have thousands of hooks. Turtles, seabirds, sharks, and other types of fish end up biting more often than tuna.

Apparently there's an odd explanation for why large shoals of bluefin tuna found their way to Vestfjorden. From time immemorial, all the way back to the age of the Phoenicians, there have been huge numbers of tuna in the Mediterranean. In Italy the fish is called *tonnara;* in Spain it's *almadraba.* Bluefin tuna spawn in the Mediterranean, and throughout history, tens of thousands were caught every year. Dense shoals of tuna were guided through a labyrinth of nets to shallow waters, where they were clubbed to death. As long as enough fish escaped and managed to make their way back to the Atlantic Ocean, the fishing was somewhat sustainable.

To make himself popular among his own people in Andalusia, the Spanish dictator Francisco Franco built a number of factories where the fish were processed and packed into millions of cans. New technology made the fishing more efficient, and a new fleet of large motorized vessels followed the tuna out into the waters of the Atlantic. The Second World War put a stop to the overfishing, and the tuna populations recovered. After the war, the Bay of Biscay was littered with mines. The Spanish and French fishermen didn't dare fish in that area. This too added to the increase in stock, and the bluefin tuna grew in huge numbers all the way up to Vestfjorden.

Yet after ten short years, the tuna disappeared from the Norwegian coastline, and for several decades it has been considered an endangered species. But during the past few years it has again been observed off the coast of Norway. The Japanese will pay more than a million Norwegian kroner for a perfect specimen. But the fish has to be alive so it can be fattened up before slaughter. The meat around its belly contains a butterlike, pleasant-tasting fat. No other raw material is as coveted in sushi restaurants. I've seen where the catch ends up: in the famed Tsukiji fish market in Tokyo. There, in vast hangarlike halls, the fish lie lined up like unidentified corpses after a plane crash, a tsunami, or some other major disaster.

Who knows? Maybe the bluefin tuna will come back to Vestfjorden, fifty years after it disappeared. Hugo has started keeping an eye out for tuna whenever he goes out to sea. Many other exotic species keep showing up along the Norwegian coast, such as the ocean sunfish, European sea bass, Peter's fish, and other "tourists." A few years ago a swordfish was caught in a net in Steigen, very close to Hugo's place. Colonies of tropical jellyfish, commonly called sea raft or by-the-wind sailor, have washed ashore in Lofoten. They float on the water and have little sails, which cause them to be blown across the oceans. No one had ever seen any in Norway before.

This sort of phenomenon is most likely due to global warming. But don't think it will make these waters richer. As new and ill-adapted species move farther north, the fish already here will slowly but surely move farther north if it gets too warm along the Norwegian coast.

At night I sleep with the window open. Only a gentle breeze ripples through the air. The muted, soft gurgling of the sea against the rocks underneath Aasjord Station finds its way through the thin veil of sleep. On the seaward side of the Vesterålen archipelago, they have a special word for the sound of the ocean when heard through a bedroom window on a mild summer night—the sound of water calmly lapping against the shore. The word is *sjybårturn.*

10

The next morning we haul out a couple of crab pots and a long fishing line with dozens of hooks. It's just as warm and calm as it was yesterday, and the dog days of summer (July 23 to August 23) have just begun. We can already sense it. Seaweed has detached from the seafloor and is now quite visible as it floats in the water. That's how the ocean regenerates.

Dead bodies that have lain on the bottom also have a tendency to rise to the surface at this time. The sea gives up its dead, as it says in Revelation 20:13. In the old days, people believed food spoiled easier when the dog days set in and that the flies were more numerous and insistent. The sea reaches its warmest temperature during the dog days. The algae bloom depletes the seafloor of nutrients and oxygen, and many jellyfish appear in the ocean. They swim or drift around, pale and yellow, like fringed moons.

As we set out the crab pots, we know they will be filled with brown crabs when we come back to get them in the evening. But are the crabs safe to eat? The level of the heavy metal cadmium is so high that the health authorities have issued a warning. We discover later that two of the crabs have ugly black patches on their shells, no doubt caused by some contagious disease. Hugo also tells me that in the past ten years very few wolffish have been caught. At first he thought it was being fished farther out to sea, in the wintertime. But on those occasions when he did catch wolffish, he noticed many of them had abscesses that looked like cancerous sores. Today the wolffish is making a comeback, though no one knows exactly why.

The sea in Vestfjorden seems cleaner than in most other places in the world. The ocean is deep, the current strong, and huge masses of water are recirculated each day. But there are more heavy metals in this area than farther south, maybe because the sea is like a giant organism, and these open waters are directly linked to the global system of currents, many of which are northbound.

At last we're going to put out our baited fishing line, five nautical miles south of the Skrova lighthouse. Hugo stays as far away as he can possibly get in our small boat as I cut open the remaining sack of offal from the slaughtered bull. The corpse stench pours out of the bag to hover over Vestfjorden. If we're lucky, we won't have a Greenland shark jumping into the boat while I'm attaching a hip joint covered with rotting red meat to the big, shiny hook. I don't know what sort of expectations the Scottish Highland bull may have had, but I'm pretty sure the animal never imagined anything like this.

We make sure we're positioned in exactly the same spot as yesterday when we sank our chum bait to lure the shark. Then I drop the hook over the side. As Rimbaud put it: "Hideous strands at the end of brown gulfs / Where giant serpents devoured by bedbugs / Fall down from gnarled trees with black scent!"[33] I let the chain and the line run down to the bottom, not stopping until the reel is almost empty, which means we've let out close to eleven hundred feet of line. The twenty feet of chain at the bottom are essential because if the Greenland shark bites, it will start to roll around. The skin of the shark is so rough that a chain is the only thing that will hold. If you were to stroke a Greenland shark away from the direction it swims, its skin would feel smooth and frictionless. But if you stroke it in the opposite direction, you would cut your hand badly, because the shark's skin is covered with tiny "skin teeth," sharp as razors. Before World War II, Greenland sharks were exported to Germany, where the skin was used for sandpaper. The liver of the shark was boiled, and the fat was used in the production of glycerin and nitroglycerin. The latter, a highly unstable explosive, would often go off accidentally when sparked by small shocks or friction, and kill the people who handled or transported it.

Finally Hugo fastens the line to our biggest float and tosses it into the water. The float is now transformed into a fishing bob, which is something I often used as a boy. But back then I was fishing for perch, brown trout, or arctic char—fish that weighed a pound at most. The bob was the size of a matchbox. You might say we're still using the same sort of gear, but now we've grown up, and we're fishing for Greenland shark with a bob that measures a yard across. And instead of a worm on a half-inch hook, we're using a hook that looks like something used in a slaughterhouse, with parts of a huge dead beast attached to it. But that's what we need. Even a Greenland shark won't be able to pull the float under, at least not for more than a second at a time.

Wanted: one medium-sized Greenland shark, ten to fifteen feet in length and weighing about thirteen hundred pounds. Latin name: *Somniosus microcephalus*. Blunt, rounded snout, cigar-shaped body, relatively small fins. Gives birth to live offspring. Lives in the North Atlantic and even swims under the floating ice cap at the North Pole. Prefers temperatures close to freezing but can also tolerate warmer water. Can dive to a depth of four thousand feet or more. The teeth in its lower jaw are as small as a saw blade's. The teeth in the upper jaw are equally sharp but significantly bigger, and are used to bore into the prey while the lower teeth saw their way through. In addition to saw-blade teeth, it has, like a few other types of shark, suctioning lips that "glue" larger prey to its mouth while chewing. And every mating act is violent. On the bright side, the Greenland shark does not have sex until it's about 150 years old.

Scientists who have examined the stomach contents of Greenland sharks have encountered many surprises. How is it possible that in Greenland, Fridtjof Nansen (1861–1930), the famed Norwegian scientist, explorer, and politician, opened the stomach of a shark he'd caught and found a whole seal, eight large cods, a ling four feet long, a big halibut head, and several chunks of whale blubber? Nansen claimed, by the way, that the shark was able to live for several days even after this "huge, ugly animal" had been cut open and placed on ice.[34]

The eye parasite *Ommatokoita elongata,* which is about two inches long, slowly devours the cornea of the Greenland shark, until it goes blind. In the folds of its belly the shark also has other parasites in the form of little yellow crabs (*Aega arctica*). Old shark fishermen have recounted how the parasites would fall off by the hundreds when the shark was hoisted aboard.

The Greenland shark can be used for more than just making sandpaper and nitroglycerin. Its flesh is poisonous, smells like urine, and can serve as a potent drug. The Inuit used to feed the meat to their dogs, if nothing else was available. But the dogs would get extremely intoxicated and might even end up paralyzed for days. During World War I, there was a shortage of food in many places in the north, and people couldn't be choosy. There was more than enough meat from Greenland sharks. But if people ate the meat when it was fresh, or neglected to treat it in the proper way, they could get "shark drunk," because the flesh contains the nerve gas trimethylamine oxide.

The resultant inebriated state is supposedly similar to taking in an extreme amount of alcohol or hallucinogenic drugs. Shark drunk people speak incoherently, see visions, stagger, and act very crazy. When they finally fall asleep, it's nearly impossible to wake them up. To avoid these side effects, you need to cut the main artery of a Greenland shark immediately, so that the blood drains out. Then the meat can be dried or boiled in water, which has to be changed several times. In Iceland, the shark (called *hákarl*) is considered a delicacy, but there everyone is careful to prepare the meat properly. To make the poisons disappear requires repeated boiling, drying, or even burying the meat until it ferments.

It should be no surprise that people living in northern Norway developed a healthy skepticism when it comes to the meat of the Greenland shark. The reason they even bothered to catch it was because the liver is extremely rich in oil. In the 1950s, Norway was the leader in commercial fishing for the Greenland shark, but by the early 1960s, demand was already fading.[35] Only now is it making a small comeback.

—

Our boat is gently bobbing in the sunshine in Vestfjorden. Yesterday the sea glittered and crackled with light. Today it has a steady, calm glow. The ocean has found its lowest resting pulse, as it does only after many days of good weather in the summertime. It's also a neap tide, which means the difference between high and low tide is unusually small. The gravitational force of the moon and the sun pull the sea in opposite directions, canceling each other out to a certain extent, like when two people arm-wrestle and neither has an advantage.

Our only task is to wait and keep an eye on the floats. Maybe it's because we're drifting in Vestfjorden—where the currents function just fine on their own even when there's no wind—that Hugo happens to think of a story about one time when he and his brother were out in their fishing smack. The boat, called the *Plingen,* was a small carvel-built vessel made in Namdalen in the 1950s. The fishing smack was waterlogged and sat low in the sea. In bad weather they had to pump out the water frantically by hand. One ice-cold day during the Lofoten fishing season in 1984, the two brothers went out during a squall. The motor wouldn't start, but another boat in the fishing grounds saw they were in trouble and towed them back to Svolvær.

That reminds Hugo of a similar situation. They were on board the *Helnessund* heading out of Svolvær after picking up a cargo of fresh shrimp that had been caught farther north in Finnmark. When a storm blew in, the boat quickly ran into trouble. The refrigeration unit failed and the cargo shifted. The freighter ended up drifting in the middle of Vestfjorden. By using countless buckets of seawater, they were finally able to cool down the engine enough to make it over to Skrova.

Hugo often makes these sorts of associative leaps. When one story starts getting a bit worn out, it taps the next one on the back and sends it off, in a relay race that can go on and on. The stories usually move further and further away from the starting point. Sometimes I get confused and wonder what Hugo's stories have to do with anything at all.

But something about what he has already told me makes Hugo think of Måløya, one of the small islands on the seaward side of Steigen. That's the location of a tiny, abandoned community that Hugo was curious about. Together with his brother, he dropped anchor and left the fishing smack to row a skiff, or *reksa,* as Hugo always calls these small wooden rowboats, toward a gently sloping sandy beach. But they misjudged the waves, and the little *reksa* got tossed around. Both brothers ended up in the icy water. They went ashore but didn't stay long because it was late winter, and the air and water were freezing. On their way back to the fishing smack, the *reksa* again filled up with water because a small crack in the bottom was now much bigger after the rowboat had been tossed by the waves. Just before the *reksa* sank, the brothers managed to grab hold of the fishing smack, not by the gunwale, but farther down. They clung to the scupper. It was impossible for them to haul themselves on board, exhausted as they were, and with their soaked clothing heavy with seawater. After hanging there for a while, side by side like in some cartoon, they both realized how absurd the situation was and burst out laughing. But their strength was about to give out, and they needed to focus all their efforts on one last-ditch attempt to save themselves. So Hugo became a human ladder for his brother to climb up and clamber on board.

If Hugo had lost his grip before his brother made it on deck, it's unlikely either of them would be here to tell the story. But Hugo seems to think the main point of the whole tale is that a person doesn't actually get all that cold by floating in Vestfjorden for nearly half an hour in March.

"We stayed out for the rest of the day, and without changing our clothes. Although, I have to admit that behind our ears, and at the back of our necks—that's where the cold settled."

Sometimes I wonder whether my friend is actually part sea mammal.

II

What's going on down there on the seafloor, more than a thousand feet below us? Is the beast starting to sniff its way toward our stinking bait? The oily substances of putrefaction must be spreading like smoke from a fire, way down there in the water. What are we going to do if we actually manage to bring the shark up to the surface? I feel a certain fear mixed with anticipation at the thought.

An acquaintance who used to be a seaman on a trawler once told me what they would do if a Greenland shark got caught in the trawl net and ended up on deck. They would tie a line around the base of its tail, lift it up with the derrick, and swing the shark out along the side of the ship. Then they would cut off the tail so the shark fell into the sea with a big splash. The amputation was accomplished in a flash, since Greenland sharks, like all sharks, have no bones, only cartilage. The shark is very much alive when it lands in the water, but it quickly discovers that something is seriously wrong. We humans wouldn't have much of a chance if someone chopped off our legs and arms, and threw us overboard in the open ocean. Without its tail fin, the shark is helpless. It can't move forward or keep its balance in the water. After a short time it sinks to the bottom, and down there in the ice-cold dark, it's most likely eaten alive by other Greenland sharks.

Hugo tells me that something similar was done with basking sharks. It was common to turn the shark over and cut open the belly so the liver floated out. Then the shark would continue swimming without its liver, at least for a while.

They didn't always cut off the tail of a Greenland shark. My friend the trawler seaman told me that sometimes they also painted the name of their boat on the side of the shark, as a greeting to the next trawler that happened to catch it. Whoever landed a shark like that in their net would paint the name of their own trawler on the other side of the fish and then set it free again. It probably would have been easier to send a postcard, but trawler crews have their own brand of humor.

"Look! Isn't the float moving?" Hugo says.

It looks like it's popping up and down with the unnatural rhythm of a gigantic fishing bob. Something is definitely happening a few hundred yards away from where we're sitting, in the middle of a shoal of mackerel. Hugo starts the motor, and in sixty seconds we're at the site.

Hugo starts pulling up, meaning he hauls on the line, and there's no doubt that something big has taken the bait. After a while I take over, and it goes even slower. Have you ever tried to haul up from the seafloor a Greenland shark that's maybe twenty-two feet long and weighs fifteen hundred pounds? A shark that's holding on to an eleven-hundred-foot line attached to twenty feet of chain? The line digs into my fingers. It's sheer agony. Stinging jellyfish have attached themselves to the line, and we're not wearing gloves.

My arms are practically lifeless, and there's no more than 150 feet left when all of a sudden it gets a lot easier. Anyone who has ever gone fishing knows the feeling of deep disappointment. In a hundredth of a second all your hopes are crushed. You go from being excited, determined, and focused to tumbling down the basement stairs. Even though the line had been cutting into my hands, it hurts more to feel the weight disappear. Hauling the line the rest of the way up seems like harder work, even though it weighs next to nothing. A few minutes later the hook attached to the chain is right under the boat. I lift it up so it dangles in front of us. When we dropped the baited hook in the water, the hip joint was covered with red flesh. Now it has been gnawed clean. Dozens of tiny orange parasites are scratching at the bone. They look like lice or little insects. They must be what live in the folds on the belly of a Greenland shark.

We can clearly see the sawlike marks left by the shark's teeth on the bone and fat. The hook is stuck through a tendon in the joint so it lies right next to the bone. I had assumed the shark would crush the whole thing if it took the bait. But it didn't. And that's why the shark wasn't hooked more firmly. That's why it was able to pull free or simply let go. That's why we're sitting here, not saying a word. I tell Hugo about my mistake, and he just gives me a pensive little nod without a hint of blame.

After the initial disappointment has subsided, we decide not to consider the episode a defeat. Instead, we will view it as a sign that we're doing things right. Not many people have been on the verge of catching a Greenland shark on their very first try. All we can do is rebait the hook and toss it back into the water.

Down there, underneath our pontoons, our monster is swimming, waiting to be fed. Several hundred yards away, toward shore, a boat is anchored. It's filled with happy youths enjoying the glorious weather. The girls jump into the water, which is cold, but it's never going to be warmer than it is right now. If they knew what was staring up at them from the deep as they splashed around, they would rush back on board. One of the girls is wearing an orange swimsuit and probably doesn't know that the colors yellow and orange seem to provoke sharks to attack. Australian divers and surfers never wear gear in those colors.

We get no more bites that day. Or the next. On the third day we let the baited line stay out there all night. The following morning the whole thing is gone, as if sunk into the sea. Both floats are probably far out in the ocean, dragged off by some invisible force, either by the current or by a Greenland shark. It would be pointless to go looking for the floats. Even if we had all the time in the world and an unlimited supply of gasoline, the chance of finding them would be close to zero.

Three days later we're on our way back across Vestfjorden. We've written off the floats and chain, along with the fishing line. But out in the middle of Vestfjorden, in poor visibility and high seas, we practically run right into them. The whole length of line and chain is there. Only the hook and shackle, the U-shaped steel clamp fastening the hook to the chain, is missing. That's incredible. It shouldn't be possible for a clamp, firmly fastened with a pair of pliers, to come loose. And it would take tremendous force to snap it in two. Yet one of these two things must have happened. At least that's what we tell ourselves. But the truth is, it was amateurish for us to leave the baited line out here overnight. Local fishermen confirm this. The current is so strong that it will take *everything*, if given enough time.

Our boat carves a white V across Vestfjorden. Out in the ocean a small rainbow forms a portal. It's tempting to set course toward it, just to drive through. But we're not hunting rainbows.

The mirage line, where the sky meets the sea, is suddenly obscured, creating optical illusions. Several small islands out there seem to be much closer, floating above the glittering sea. The sun is burning the edges of magnesium-white clouds far to the west. It has rained, and we can see individually delineated local showers pouring down way off in the distance. The sun isn't visible to us, but it casts its light around and in between the rain, in some places looking like gigantic spotlights slowly sweeping across the surface of the water. Where we are, the world seems cleansed and filled with mirrors. The colors are oyster shell and slate.

12

As we approach Engeløya, great shoals of Atlantic mackerel splash around us, no doubt feeding on zooplankton. Hugo isn't interested, and he snorts scornfully when I suggest catching a few to put on the barbecue. Like so many other northerners in Norway, he detests this type of fish. He can't stand the taste. He has tried to prepare mackerel in numerous ways, but it hasn't made any difference, no matter what recipe he uses. So far Hugo hasn't discovered any surefire method for removing the mackerel taste from mackerel. He says I'm welcome to catch a few to throw on the barbecue later, as long as he doesn't have to be anywhere near.

The northerners' contempt for mackerel has a long tradition. People believed this fish, with a pattern on its back that looks like a human skeleton, ate the bodies of the drowned. Even further back in time, it was thought the fish also ate people who were alive. Erik Pontoppidan (1698–1764), who was the bishop of Bergen, speaks of the mackerel as a type of Nordic piranha. "Like the shark," the bishop writes, the mackerel shares the propensity for "eating human flesh, seeking out those who swim in the nude, so the person will be quickly devoured if he falls in a school or shoal of mackerel." To underscore this claim, Pontoppidan recounts a "deplorable incident" when a sailor, perhaps sweaty after hard physical labor, wanted to take a dip in Laurkulen harbor (present-day Larkollen, south of Moss in southern Norway). Suddenly the merry sailor disappeared, as if something had yanked him under. After a couple of minutes he resurfaced, his body "bloody and gnawed, and swarming with mackerel that refused to be chased away." If the man's comrades had not come to his rescue, the sailor would have "without a doubt" suffered a very "painful death," Pontoppidan asserts.[36]

In sheltered waters we stop between the islands of Lauvøya and Angerøya. There we catch a small cod, throw it back into the water, and then sit down to wait for the eagle, which has a nest at the top of the mountain, to come and sink its claws into the half-dead fish floating on the surface. We can see the eagle, but it doesn't dive for our bait the way it has done so many times before. A seagull flies over to us. Its body looks smaller than the cod's, but with effort the gull swallows the fish whole. The bird's belly is now so full it fails to take flight. Sometimes we bite off more than we can chew.

Autumn

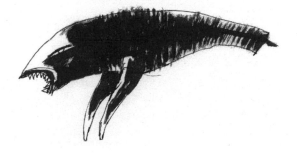

13

The next time I fly north, the birds have migrated in the opposite direction. It's early October, and a silence has settled over
the land. The trees, bushes, and plants are retreating toward
their roots, getting ready to go dormant before the snow and
frost arrive. A heavy, dark tone hovers over inland Norway. The
lakes will soon turn white, and the valleys will fill with snow.
Near the shore, and in the sea, it's a different story. There life
reawakens when the water gets colder and storms whip up the
waves. The crabs move faster, the flounder get more brazen, the
pollock are firmer, the shellfish taste better. We're approaching
the winter fishing season in northern Norway.

Again Hugo and I cross Vestfjorden from Steigen to Skrova.
This time the sea is black as ink and possessed of a restless agi-
tation. The light has dimmed, the cloud cover reaches down
almost to the water. Hugo navigates in a zigzag pattern, meeting
the waves from the side or the back, so he can surf as much as
possible. In spite of the *unnafuringen,* which is what Norwegian
fishermen call this maneuver when they swerve to avoid the
worst of the head waves, it's an unpleasant crossing.

As we approach Skrova, Vestfjorden shows us a little of its
power. The sea is cold and raw, the rain lashes the whitecaps,
which crash onto the shore with muted roars. The sea and the
sky are not calm, separate entities, as they were the last time we
were here. Today they form one continuous, churning motion.

The Lofoten Wall isn't visible until we're only a few nautical miles away. Hugo steers the boat between the skerries and islands to enter Skrova harbor.

The bad weather continues over the next few days, keeping us from going out to sea. Instead, I help Hugo with some of his chores, which tend to pile up when you're in charge of numerous wooden buildings encompassing tens of thousands of square feet.

Aasjord Station consists of two large buildings. The main building, which stands on the seaward side, has three stories totaling at least ten thousand square feet. Behind the main building is another building almost equal in size, and it too has three stories. Next to it stands a one-story cutting shed. The main buildings have been used as a fish landing center, cod-liver-oil mill, saltery, fishing gear storeroom, and dried fish warehouse.

The three buildings are connected, almost in the same way as the Holy Trinity is—meaning they are one being and yet individual entities. Inside, you hardly notice when you move from one structure to another. After visiting so many times, you'd think I'd know every nook and cranny, but that's not the case. Every time I leave the beaten track inside the main building, I discover some room or attic space, maybe even entire sections, that I've never seen before. It's as if the fishing station is inexhaustible and will always have an undiscovered room to offer. And by the way, the island seems to be the same way. Each time I take a walk on Skrova, I end up in places I didn't know, discovering a new sandy beach or an old German bunker atop an inaccessible hill.

Behind the fishing station, up on the steep slopes, there are two other small buildings: the Red House (Rødhuset) and the White House (Kvithuset). To help out, I stack up posts that have been used for the fish-drying racks on the slope. In the meantime, Hugo does some carpentry work on the Red House. Hugo and Mette are planning to move into the Red House when it's ready. Spending the winters in the huge, drafty main building holds no appeal. Hugo is installing insulation in the Red House, making a thorough job of it. He is also replacing the walls, the roof, and the floors. At the back he's building an annex that will be their bathroom.

He has already finished working on the White House. It's an authentic old fisherman's house from the early nineteenth century and considerably older than the fishing station. Hugo grabbed this house from the jaws of demolition a few decades back, before he even owned Aasjord Station. Now he has put up new board siding, replaced the windows, added insulation, put on a tar paper roof, built steps and a porch, and installed an old wood stove. From the windows on both the ground floor and upstairs, you have an unobstructed view of the bay. He used vintage glass, which gives the outdoors a blurry, distorted, and vaguely dreamy appearance, almost like being under water. When he pulled off the old paneling on the second floor, he discovered the walls had been insulated with newspapers from 1887. He decided to preserve them by coating them with varnish.

Instead of telling Hugo how impressed I am, I take on the role of a pedantic building inspector as he shows me around inside the White House. I walk with my hands clasped behind my back as I ask him why he did such and such when it might have been better, maybe even smarter or more in line with regulations, to do things some other way. It takes Hugo a couple of minutes to realize I'm joking, and then he sends me off to carry posts.

Since I'm not of much practical use at Aasjord Station, I find it best not to squander a good chore like this by finishing it too quickly. Better to make it last for a while so I'll feel like I'm being handy.

After a short while I go inside the station, and before I know it, I've stumbled upon a room I've never seen before. On a shelf I find a bunch of newspapers, yellow with age. I pick up one of them, lean against a windowsill, and start to read an issue of *Nordlands Framtid* from September 8, 1963. At the top of the densely typeset front page, it says: "Norwegian naval vessels bomb Å in Lofoten with high-explosive shells." That sort of headline can't help but pique my interest, so I keep reading:

During firing practice carried out by naval vessels near Lofoten on Sunday, an error led to a number of shells falling on the community of Å on the island of Moskenesøya. It's a miracle that no one was killed or seriously injured. A shell hit an outbuilding in the middle of the village and exploded, and shrapnel penetrated halfway through the timbered side of a residence five meters away from where the family was eating dinner. Twelve to fifteen other shells swooped right over the heads of people in the little fishing village, and many had to dive for cover in the ditches while the onslaught lasted. Four shells struck the village itself, while eight others landed among the fishing boats in the harbor. When the outbuilding exploded, three ten-year-old girls were walking past on the main road just 15 meters away. They were only slightly injured by shrapnel, which scattered in a 50-meter radius. Lamps and book-shelves in the nearby home fell off the walls, and a table toppled over in the living room. Less than 30 meters from the site of the explosion, five cabs carrying about 20 tour-ists had just stopped to take in the view, but no one was hit by shrapnel.

The sheriff was immediately notified, and the telegra-pher at Sørvågen radio finally made direct contact with the destroyer *Bergen* and was able to stop the bombard-ment before any lives were lost.

That's the Norwegian navy for you. They had huge areas devoid of people at their disposal, yet they still managed to fire shells into the middle of the little fishing village of Å, which is surrounded by wilderness at the remote western tip of Lofoten. It had to be an accident. If the navy had actually tried to hit that small target, they probably wouldn't have been able to.

A copy of the newspaper *Nordlandsposten* from January 24, 1964, also contains dramatic news. In a lengthy letter to the editor, under the headline "Broomstick Murder," Halvdan Orø takes to task a man who used a broomstick to kill an otter. "A broomstick of such poor quality that it breaks in half during the killing is not a proper weapon, and one has to ask whether the death of this otter might be characterized as animal abuse."

With the help of distractions like this, the chore of carrying posts takes me a disproportionately long time. When I'm finally finished, I feel as if I've pretty much done my part. I've done more reading than lifting, and there's nothing wrong with my muscles, but Hugo's the one who has to tell me how to make best use of them. Hugo scratches his head but can't come up with anything, so he lets me off the hook. Otherwise I'll just get in the way and delay his own work. To be honest, that suits me fine. I've brought along to Skrova a bunch of old books that are of special interest to anyone who wants to know about the sea. This time the magnum opus I'm reading is Olaus Magnus's voluminous work from 1555, originally written in Latin: *Historia de gentibus septentrionalibus* (*A Description of the Northern Peoples*).

14

With each day the sea gets more restless. The barometer falls. Out on Vestfjorden a stiff wind whips at the crests of the roiling swells and whitecaps, lifting the water into microscopic drops that fly through the air. From a distance it looks as if smoke is coming off the sea.

Black clouds hover low in the sky, but occasionally openings appear. Then rays of translucent sunlight fall over the island, illuminating and magnifying everything they touch. Sometimes Aasjord Station is dazzling white. Other times it's as gray as the skeleton of a beached whale.

Then the rain arrives. A heavy, monotonous, dreary rain with no end in sight.

Gusts of wind from the west bring the rain. Everywhere in the world the oceans, promontories, straits, islands, and coastal areas are dominated by particular winds. On Vestfjorden, like most of the sea areas in the Northern Hemisphere, the west wind is prevalent. The scientific explanation is that high pressure over the Azores and low pressure over Iceland produce a strong west wind in the North Atlantic.

On old maps the winds were often depicted as having faces. Maybe this was a tradition from antiquity, since the Greek gods associated with natural phenomena led very busy lives. Aeolus, god of the wind, was the son of Poseidon, god of the sea.[1] His bulging cheeks were usually his most prominent feature, and he blew the west wind with all his might.

Back when all ships were at the mercy of the weather, the winds were ascribed certain traits, even personalities. Some winds were sly and capricious, but luckily there were people who knew how to control them. In the mid-1600s, the Frenchman Pierre Martin de la Martinière was sailing north as captain of a large ship. The wind subsided, and the ship, having reached an area south of Lofoten but north of the Arctic Circle—near Bodø, in other words—wouldn't budge. The captain made contact with the local wind conjurers, "children of the wind's prince," who were said to be able to summon both storms and calm weather—for a price. A wind conjurer came out to the sailing ship and instructed the crew to tie three knots in a woolen cloth and then fasten it to the ship's foremast. Whenever they needed wind, all they had to do was untie one of the knots. La Martinière was highly skeptical, but as soon as the first knot was loosened, a brisk southwesterly wind filled the sails, taking the ship farther north.[2]

Today, meteorologists generally use eight different wind directions: north, northeast, northwest, south, southeast, southwest, west, east. In the old days, the wind was divided into sixteen different directions, roughly speaking. Arthur Brox, from Senja, a large island north of Lofoten, recorded thirty local words for various types of wind.[3]

Some wind terms included details about how the landscape and the wind work together. For example, if there was a strong wind blowing from the south, it was of interest to know how the southern wind would strike a specific local area. Was it a *landsønning*—a south wind coming from land, which on the northern coast would mean from the southeast? Or was it an *utsønning,* a more cunning variant for seamen, since it arose out at sea?

A southwesterly wind is the worst kind on Vestfjorden.

For the most part the buildings of Aasjord Station have no insulation and breathe with the wind and weather. In a strange way they also seem to be pervaded by everything and everyone who has ever been here. Operations ceased in the early 1980s, so you need to have a keen sense of smell to detect any trace of the millions of fish that have passed through these halls. Yet there are many other lingering traces, as if the buildings themselves have memories and are capable of conveying a vague impression of their own past, surreptitiously and imperceptibly, the way rumors sometimes appear in dreams.

Maybe this has something to do with all the abandoned objects. Almost everything is still here from when the fishing station was operating. Aside from small items that were taken away, and others that may have been stolen over the years, most things remain exactly where they were left. Many tons of heavy nets and coils of rope are piled in the corners. Wooden brine tubs are still filled with salt. A glazed crust has formed on top, but all you have to do is punch a hole through it to get to the salt. The Aasjords have enough salt on hand to last for ten generations or so.

In many of the small rooms I see work clothes hanging on hooks, as if the next shift will soon come in the door. But the clothes probably belonged instead to the crews of fishing boats, most of them beached long ago. And the workers themselves who were young the last time they were here are either old or dead by now. Other personal belongings, kitchen equipment, and delivery slips for fish are scattered around the former living quarters. In an old office, there are even completed delivery reports hanging on the wall. They show a record of how many stockfish the station bought during the first three months of 1961 (112,727 kilos, almost 250,000 pounds), how much was in production, how much was sold and delivered, et cetera. The chart has a separate column for "Goods Sent to Bergen."

All the station's products were meticulously recorded: raw fish, salt fish, dried fish (of various qualities), liver (untreated, preserved in alcohol, or boiled), all sorts of oils (centrifuged cod liver oil, hot-pressed oil, sour oil, ind. pressed oil, and finally: "other oils"). The report continues with roe (raw, sugar salted), salted scrap fish, fish heads, and at the very bottom of the form: liver *graks,* meaning the waste left over after steaming the cod livers.

Many years of work haunt this building, from the moment the first nails were pounded into the boards until the last tenant left the place. The fishing station is a marinade of memories. I imagine invisible clocks hanging on the walls in the various sections, all of them showing different times. None of them shows the current time; most stopped working decades ago.

—

In the 1980s, the fishing station was bought by two Finns, and they also left traces behind. They are still alive, somewhere in Finland. Her name is Pirrka, and his is Pekka. She's a renowned psychologist, and he's a documentary filmmaker. In the 1970s, he made ethnographic films in remote countries, and many of his films have achieved cult status. Two educated and cultivated Finns who spoke quietly, thoughtfully, and only sporadically, at least on those occasions when I've met them. In fact, they spoke as if they were in a sauna, even when they were freezing cold—and that happened often in Skrova. He was interested in flowers, and there are lots of them in an amazingly warm and lush valley out toward Hattvika, in the middle of Skrova. If you head in that direction, you don't expect to see much more than rocks and crags, maybe some gullies and ravines. But suddenly you find yourself in the center of a glade.

In the rooms where Pirrka and Pekka lived, there are still big stacks of the Finnish newspapers *Hufvudstadsbladet* and *Ilta-Sanomat.* Hanging on the wall is a satellite photo of the Finnish-Swedish archipelago, which is called *saaristomaailma* in Finnish. There are thousands of tightly packed little islands in a wide belt between Finland and Sweden. A gap of about twelve miles is what makes it possible for ships to slip into the Gulf of Bothnia.

Only the gods know how Pirrka and Pekka ever found their way to Skrova. But they fell in love with the place and bought Aasjord Station when they were traveling north and happened to hear it was for sale. Neither of them was young. Each summer they spent a few weeks of vacation here. They lived in a small corner of one building—like dethroned aristocrats who had lost all their money and titles but insisted on barricading themselves inside a tiny space in their dilapidated castle. Even though they clearly were infatuated with Skrova and Aasjord Station, they seemed overwhelmed and a bit out of place in these surroundings. Someone in their family or circle of friends must have been into diving (the rubber boat they used, now punctured, is still lying outside), and maybe that was who persuaded the couple to buy a huge fishing station on a little island in Lofoten. Every summer lots of Finns left their green coasts and lovely lakes to visit Skrova and go diving with wetsuits, flippers, belts of lead, and harpoon guns. Most of their gear still hangs on hooks in the station. Pekka and Pirrka didn't go diving, but they were definitely in over their heads.

In the end, they let Aasjord Station return to the Aasjord family. It has been fifteen years since Pekka and Pirrka left Skrova. Hugo talks about them as if they might show up at any minute, but that's not likely to happen.

One afternoon when the weather is too bad for us to go out fishing, Hugo and I end up in the attic. The space is filled from floor to ceiling with old gear. You could actually start your own fishing station and cod-liver-oil mill with all this equipment, which is more than one hundred years old. We find boilers, presses, oil vats, separators, pipes, grindstones, net floats, and Bismar scales; hoisting devices with pulleys, gears, and winches, huge washing vessels made of wood, electrical motors, landing nets with shafts many yards in length, herring landing nets, and mysterious tools made of wood and metal. In one room there are dozens of oak barrels for cod liver oil. Some are stamped "medicinal oil," others "sour oil." Several smaller barrels may have once held cognac, since smuggling is commonplace along all coasts. The purse seiner *Seto,* for instance—which doubled as a freighter that made trips to the continent in the off-season— was notorious for its smuggling activities, which were well known even to the customs officials.

The technology on display in the attic is simultaneously out-dated and advanced. Much of it was made on-site and developed over hundreds of years by mechanics, coopers, carpenters, smiths, rope makers, and local self-taught smart guys who could solve any problem using whatever they had on hand in terms of materials and tools. But as far as I'm concerned, much of it is highly puzzling. I point at a strange little device with a steel sluice at the top. Obviously something was meant to go in one end and come out the other. It looks as if it should be attached to some other piece of machinery.

"That thing over there? It's a scale scraper for pollock, of course," says Hugo as he keeps walking.

"Oh, sure. Of course. It's not a haddock scraper, it's a pollock scraper. I couldn't really tell in this dim light," I reply.

Hugo looks at me and laughs.

Every time we discover another obscure object, Hugo starts what I can only describe as sparring with it. He dances around it, jabbing out his hands at varying heights as he speculates about the specific object's form and function. In his attempts to wrest the secret of its use, he might point to where something could be inserted or ejected, or focus on something that's supposed to rotate and determine which way it should be turned, or notice how one part interlocks with another, and so on. Finally he comes up with a theory he's happy with, and it usually sounds convincing to me too.

Now, I see to my satisfaction that he has come to a standstill at the very back of the attic. He's looking at some sort of cast-metal pulley with two wheels, a handle, and several protrusions made of steel. The whole gadget is five feet long and barely reaches to our knees.

"Give me a week and I'll figure it out," he says.

"You have twenty-four hours," I tell him.

Hugo sometimes seems like an absentminded professor, and there's no doubt he could have created strange mechanical devices from all the junk up in the attic. Machines for utilitarian purposes as yet undiscovered. Motors powered by electric eels and lubricated with oil from a Greenland shark.

But first we need to catch the shark, of course.

15

In the evening we occasionally watch TV, always turning to one of the animal channels. They apparently broadcast nonstop programs about whales and sharks, complete with darkly dramatic music, and a lot of things depicted as dangerous! bestial! monstrous!—especially when sharks are involved. The animals are presented in an almost medieval way, judged as moral or amoral creatures whose thought processes are more or less the same as our own. The whales are mostly nice, almost bourgeois, with their nuclear families, their songs, games, and child rearing at the center of their lives, between mostly vegetarian meals and vacation-like trips through the oceans of the world.

But once in a while clips are shown that break with the pattern. In one program, a female free diver tries to befriend a pilot whale. The whale grabs hold of her foot and drags her down at least thirty feet, which is deep for a human being. There it lets go, allowing her to return to the surface, where she gasps for air. Then the whale pulls her under again and holds on until she almost drowns. The whale is not biting her, just keeping a firm but careful hold on her foot. The animal seems to be playing with the woman's life. After several more times up and down in the water column, her movements start to get lethargic. She's about to lose consciousness. The pilot whale clearly senses exactly how much the woman can take, and when she's half dead, it shoves her up to the surface. She is rescued by the same whale that almost drowned her. A good whale/bad whale scenario played out by one creature.

There's no real moral to the story other than to surmise that whales are intelligent animals and they don't automatically feel kindness or empathy toward humans. They do whatever suits them. Like all intelligent creatures, it's possible for whales to display deviant, if not exactly psychotic, behavior.

After four days I wake up with a feeling that something's not right. I lie in bed for a moment, wondering what it could be before I finally figure it out. The wind is no longer slamming rain against the outside wall. It's utterly quiet. Hugo is up and has already been out to inflate the boat.

"Ready to go out and feed the shark?" I ask.

"First we've got to get some bait, then we can feed the shark," replies Hugo.

This time, lacking a Scottish Highland bull, we're going to use some whale blubber, waiting for us at the Ellingsen facility, just across from Aasjord Station. We zip across Skrovkeila sound, which is barely three hundred feet wide, and pick up the box of whale blubber plus a trash bag containing four salmon. All free of charge. The box holds fifty pounds of blubber, cut from the belly of a minke whale. It was frozen immediately, then taken out of the freezer two days ago, but there is still some ice on it. The skin is as white as chalk. The whole piece of blubber is shaped like an accordion, except that each fold is rectangular. The surface is so smooth, elastic, and strong that it looks like something NASA would be proud to have made. The smell is pleasant. In some ways it looks like an oversized piece of bacon, clean and appetizing. Compared to the Scottish bull carcass, working with the blubber is a dream. The Japanese consider whale blubber a delicacy and eat it raw. It wouldn't take much for me to get hungry enough to fry the blubber and eat it myself.

We're going to fasten the blubber to the hook and draw the shark closer; we're going to use the salmon as chum. The fish aren't pretty enough for the European market, or maybe they were suffering from one of the many diseases that occur in farmed salmon pens. It doesn't really matter because a Greenland shark won't be picky.

We roar out of the bay and head for the far side of the Skrova lighthouse, where we fling the perforated sack with the salmon into the water. In Vestfjorden it's *opplætt*, as the fishermen say—meaning the lull that gradually sets in after a storm. Even though there's hardly any wind, it takes time for the sea to calm down. And for all we know, the storm might still be raging far out at sea.

Just for the heck of it, we also drop in the baited line, with eleven hundred feet of line, twenty feet of chain, and a hook with a thick piece of whale blubber attached. We realize there's not much chance of attracting a shark today, since the smell of the salmon hasn't had time to spread very far. But it can't hurt to try, and we need an excuse for spending several hours out in Vestfjorden.

It's raining, but the downpour has a soothing effect on both the sea and our eyes. The water is still, and each individual raindrop can be seen clearly on the surface, which is oily and smooth. If you let your eyes scan the ocean under such conditions, you can take in just about everything.

We make a quick trip over to Svolvær, where we buy newspapers and a cardboard carton of red wine, and then we stop at a café for sandwiches. Afterward we go back the same way we came and position ourselves just off the Skrova lighthouse. As expected, nothing has happened with the floats. The rain has now stopped. The sea is as smooth as a small inland lake. We read the papers and chat a bit before we head closer to shore, to the back side of the Flæsa islet, to see if we can catch a halibut or at least a cod or pollock on the line. On our way we witness a strange phenomenon. In the middle of the quiet ocean, about five hundred feet away, a huge wave is rising. It quickly grows to many feet in height, and it's coming toward us. We calmly retreat a short distance. If we'd had wetsuits and boards, we could have surfed that wave. Okay, maybe not us, but someone with skill could have done it. Then the same thing happens again. A huge wave rises up from the smooth water far from land. I look at Hugo. We've spent days out in these waters together, but we've never observed anything like this before.

"There must be a shoal there that forces the water upward at great speed when the currents are right," he says.

The sun starts to come out. The whole sea seems polished shiny from the rain, shimmering with an even gray glow. We catch nothing but a few small pollock, which we throw back. So we head back to our original position, hoping to see the floats bob under.

Hugo has a new theory about how the Greenland shark manages to catch fish and animals that are considerably faster. His idea zeroes in on a couple of aspects of the shark's anatomy.

"The lunging speed is mostly in the head or jaw, not so much in its body," he says. "The Greenland shark floats through the water, looking all innocent. If anything gets close, the shark shoves its jaw forward. The jaw isn't hinged in place, like ours. It's more like a rail or track or the breechblock of a gun."

On one of the animal channels on TV Hugo once saw something that he thinks illustrates his point quite well. The footage showed a scuba diving instructor as he approached a small shark in shallow tropical waters. Oozing with self-confidence and thinking himself master of the situation, he decided to impress the tourists he had taken along for the dive. With the camera rolling, the diver slowly glided toward the seemingly harmless shark until they were face-to-face. As the diver tried to kiss the shark on the mouth, it suddenly struck and bit off a chunk from the guy's mouth and cheek. The whole thing happened so fast that it was impossible to tell what actually took place until you played it back in slow motion. After the attack, the shark swam off among the coral while the tourists took care of the diver, suddenly in need of major plastic surgery.

"The Greenland shark has *exactly* the same kind of jaw," Hugo goes on.

There may be some truth to his theory, but it doesn't explain everything. For instance, why would a salmon get so close to a Greenland shark? And how does the shark catch big wolffish, pollock, and haddock, which all swim much faster?

"The Greenland shark is cigar-shaped, and its tail is as powerful as a great white shark's. It can use its tail to bore into whale carcasses, for example. It has the power and everything else it needs to move fast," Hugo concludes.

The hours pass. Neither of us has any complaints, and I have no wish to be anywhere else. The landscape is not in front of me. It's all *around* me. It's not something I have to pass through and put behind me. There is a strong sense of *here* in the physical ocean current near the Skrova lighthouse. It feels very far away from the information current of everyday life in which we usually float.

I'm semi-reclining in the bow, looking up. We've already drifted a thousand feet away from the floats, but they're still well within sight. Only a few faint waves roll in from open waters.

The days are growing short, the dark time is only weeks away. A few stars are dimly starting to appear in the north and east; slowly they drift in on the shoreless ocean overhead. I glimpse the contours of a few constellations. But Stella Polaris already shines brightly, its glow so wide that at first Hugo and I think it's a plane, a weather balloon, or some other unidentified flying object. It looks like the type of exaggerated drawings of the Star of Bethlehem that appear in religious literature. The star points the way to safe harbor for the two wise men in the boat.

—

Wanting a better overview, I get out my cell phone. I've downloaded an app that uses the camera and the built-in GPS to identify hundreds of constellations above us, or on the back side of the earth, if that happens to be of greater interest.

All cultures, even in prehistoric times, have seen patterns in the starry heavens, frequently naming them for gods or creatures from their own mythology. The names we use today largely stem from the Greeks, who created intricate stories about most of the constellations they discovered. (Of course no one really "discovered" the constellations, because the constellations are unadulterated products of the human imagination.) For instance, Orion is not really a giant chasing the seven virgins in the constellation Pleiades around the heavenly vault. The Greeks didn't believe that either. For them the sky was instead a canvas on which they projected their own stories.

It was a scientific activity, at least in a way, because it was about pattern recognition—something that is essential to science. For fishermen, it was basic science not only to read the ocean, the weather, and the sky, but also to remember and connect the complex patterns that arise. Only through systematic observation done over a long period of time—and making use of an innate cerebral talent—does a person get good at it.

The advent of the almanac provided fishermen with a secret weapon, because the currents, and consequently all life in the ocean, is affected to a significant degree by the phases of the moon. When the moon and the seas grow, there is more water, more currents in the fjord. And that impacts the patterns in which the fish move. For example, many fishermen knew they had to be at specific places when the moon was full in order to catch herring. If they arrived a few hours late, the herring would be gone until the next full moon.

In the old days, before fishermen had tools like GPS, sonar, echo sounders, cell phones, and reliable weather forecasts, the most skilled captains and fishermen were as respected as prominent scientists are today—at least in their local environs.

Unfortunately, the vocabulary, which was previously so rich in describing the nuances of nature, has severely diminished over the past decades. As the words disappear, so does the knowledge of complex ecological connections. Our view of the various landscapes is reduced, we attach less meaning to them, and they become less valuable to us. And that also makes them easier to destroy, in our pursuit of short-term gains.

Before long, Hugo and I will have to pull up the line and head for Skrova. But neither of us wants to discuss it. We're enjoying the silence. Our thoughts have slipped their moorings and are drifting with the current. The stars up there, the sea down here. The stars rippling, the sea gleaming and glinting.

From outer space the Gulf Stream looks like the Milky Way. From earth the Milky Way looks like the Gulf Stream. Both contain spiraling maelstroms of movement. In science fiction stories, spaceships don't look like planes; they look like boats. And they're constantly running into nebulae, ion storms, hurricanes, or icebergs. The captain stands on the bridge and looks across the deck, his face furrowed with concern. Will they make it? If the spaceship is too severely damaged, the crew will have to launch the rescue capsules, just like seamen leaving their mother ship for their lifeboats. Even the monsters in outer space often resemble creatures that are found in the sea.

Today, scientists are busy planning new types of space probes. The problem with the old ones is that they run out of power. The new ones will have tall masts with big sun sails; they will most closely resemble the old schooners or full riggers on their way through space.

I have a flat stone in my pocket. I stand up to skip it across the water. In Norwegian, it's called a flounder stone. As kids, we used to compete to see whose stone would skip the most times. If the stone is too light or flat, it spins in the air and then sinks straight to the bottom. If it's too heavy or round, it won't skip properly. I suppose your throwing technique is also a factor. I toss my stone, and it skips five times. Pretty lousy. A flounder stone probably works best on a calm body of freshwater, where you can get it to skip twenty times or more.

The rings in the water, one after another, are absorbed by the eye, which is also circular and bathed in a thin layer of salt water. Our eyes are advanced optical instruments, but the "technology" they comprise was developed over millions of years, by species that used their eyes to see underwater. Human beings can see only a very limited spectrum of light. We're not able to see many types of light waves or rays, such as gamma rays, X-rays, and ultraviolet rays. If we could, the world would look very different to us. We see with the eyes we have, and they've served us well. With the naked eye we can make out tiny plankton at close range, but we can also see stars that are thousands of light-years away, which may have burned out thousands of years ago. Lots of people have multicolored irises. If you look closely, the iris resembles a nebula. Sometimes they are composed of many colors like a galaxy, or an ocean current seen from space, scaled down and miniaturized. There is also an infinite depth to them, which can be magnified and in turn magnified again, just as the increasingly more advanced telescopes have made it possible for us to see farther into outer space.

The Greeks believed that the earth was encircled by the world-stream, called Oceanus, which was also the source of all freshwater. The god Oceanus, depicted with the head of an ox and a fish tail, ruled the movement of the heavenly bodies, which rose and fell on the horizon. And in most of Greece, that meant on the sea. After the battle of the Titans, the losers were thrown into Oceanus, condemned to drift for all eternity.

In early Greek mythology, Oceanus was a god of the heavens. A few hundred years later, after the Greeks discovered new parts of the world while they were exploring the Atlantic Ocean, the Indian Ocean, and the North Sea, Oceanus was transformed into a god of the sea. He personified the earth's oceans and was depicted with horns made of crab claws. He was frequently shown with an oar, a fishing net, and a large serpent.

"Water and meditation are wedded for ever," Herman Melville once wrote. The steady small waves lapping against the RIB's rubber sides rock us into a floating, almost trancelike state.

Where did all the water actually come from? A lot of it came from the comets that collided with our planet in its infancy in the form of ice from the cold, far reaches of the solar system, before the sun and the other planets were fully formed.

"Dirty snowballs" of rock, dust, and ice still race around in space. They're the remnants of the matter that built our solar system, back when it was mostly just flying around, colliding, collapsing, melting, and evaporating, in a continuous nuclear

Ragnarok that lasted billions of years, until the cosmic subatomic outrage eased up a bit. Gradually the solar system became more or less stable, with planets big and small falling into orbits. A few built an atmosphere from vapor and gases that were expelled from the planet's interior.

More than four billion years ago, before the oceans were formed and while the earth was covered with seas of flowing magma, the planet was bombarded by objects from outer space. One collision was so violent that big pieces of earth were blasted loose and hurled into space. Some of the pieces began orbiting around the planet. One piece that stayed is what we know as the moon. About five hundred million years ago, the earth spun considerably faster on its axis than it does today, and the moon was closer. A day lasted 21 hours, and there were 417 days in a year. At just about the same time, enough oxygen had been produced on earth to keep a flame burning. Five hundred million years—before Christ, that is.

Over billions of years a complex web of life has evolved on earth, though sometimes the actions of humans can seem quite primitive. Hugo and I, for instance, have bought eleven hundred feet of sturdy line and chain and multiple shark hooks, which we bait with a big piece of whale blubber and toss into the sea to catch a big fish for which we have no real use. At the same time, we, as a species, are capable of sending a probe far into outer space.

It took ten years for the space probe Rosetta to make it 350,000,000 miles away from earth. There it encountered the comet 67P, which is shaped like a rubber ducky and races through space at a speed of six thousand miles per hour. Rosetta sent out the small lander module Philae, which attached itself to the comet. The goal was to send analyses of the comet's water back to earth, since many of the world's leading natural scientists have been wondering how much of the water on earth actually originates from space. One theory is that not long after the earth was formed, it lost its atmosphere. The gases separating us and space disappeared. But comets filled with water and other particles bombarded the earth and thus created a new atmosphere.[4]

Unfortunately, Philae landed at an angle so that the solar cells could no longer charge, but some data were sent to earth before the batteries ran out. And many months later, in June 2015, the probe rewoke and briefly sent messages to earth.

Hugo is wearing his headset to listen to the radio. As far as I can tell he too is enjoying the situation and is in no hurry to pack up and go back to Skrova. I wave at him, and when he takes off the headset, I ask him if he knows why water exists in the universe. He smiles, shakes his head, and puts the headset back on. He probably thinks I'm joking.

It's actually not all that difficult to come up with a concrete answer to the question. The only reason that water exists in the universe is because hydrogen binds with oxygen. Swirling around the nucleus of the oxygen atom is an outer orbit of six negatively charged electrons. But the orbit has room for two more electrons. And there is one perfect partner to provide them: the hydrogen atom. Hydrogen and oxygen form covalent bonds and create H_2O. The water molecule.

Hydrogen bonds hold multiple water molecules together in a loose arrangement in which each molecule is constantly joining with the others, in a sort of dance, with the partners changing several billion times a second.

The molecules combine at a dizzying speed in ever-new variations, like letters joining together to form new words, which then become sentences and maybe even whole books. If you think of water molecules as letters, you could say that the sea contains all the books ever written in both known and unknown languages. In the oceans other languages and alphabets also arose, such as RNA and DNA, molecules in which genes connect and disconnect in the waves that wash through the helical structures and determine whether the result will be a flower, fish, starfish, firefly, or human being.

A gentle wind blows in from the rich library of the sea. The light overhead filters through the clouds, and when the rays shoot down into the water, they are conjugated like irregular verbs.

There is an enormous amount of water in outer space. But in our own solar system, water—in liquid form—presumably exists only on our planet.[5] The earth is the perfect distance from the sun. If we had been farther out in the solar system, all our water would have been in the form of ice or vapor, as in the spermlike tails of comets racing away from the sun.

The earth is big enough for gravity to hold the atmosphere in place, even though that's not a given. And we're not close enough to any giant planet with so much gravitational pull that every flow tide would make a wave several hundred feet high wash over the whole planet, like in the movie *Interstellar*. On Neptune, harsh conditions prevail. Icy winds blowing at more than twelve hundred miles per hour are constantly sweeping across the planet's polished white surface. The average temperature is about minus 350 degrees Fahrenheit. The distance from our earth to the sun is such that most of our water is liquid. Without these conditions, the water would be ice or gas, if it was present at all. And life as we know it wouldn't be able to exist.

From the boat, we see more and more stars appear on the dark blue horizon above the mountains in the east. Galaxies and planets race frictionless through space, heading farther and farther out, in an explosion that never ends. Their speed never slows. No, they're actually accelerating, though the astrophysicists don't know why. The cause lies in what they call "dark energy," which is merely a code for something they can't explain, even if most energy in the universe is of the dark kind. For whatever reason, the stars farthest away from us are picking up speed. This means that somewhere, millions of light-years away, a cosmic blind is being pulled down. Everything beyond that point is sunk in the darkness, at the bottom of a sea of stars that will forever remain unknown to everyone here on earth.

It's getting late. The moon is now in full view, and we wouldn't be able to see the floats if we didn't know where to look. I can just barely make them out. They're bobbing in the same place as we drift away, moving at a speed of several knots. Hugo seems deeply immersed in the radio program he's listening to, or maybe he's just lost in thought. I'm not about to suggest that we head back.

Moonlight takes more than a second to reach earth. Sunlight takes eight minutes. It occurs to me that astronomers are archaeologists or geologists, searching for fossils of light. Nothing happens in real time; everything we see is from the past. We're always lagging slightly behind. Even in our interactions, even inside our own heads, we're a millionth of a second behind.

Our own Milky Way, which is one among billions of galaxies, is a hundred thousand *light-years* in diameter. The most distant galaxy discovered by the Hubble telescope is a deep red patch with the prosaic name UDFj-39546284. The light from that galaxy has taken many billions of years to reach earth. The whole galaxy may have gone cold and dead billions of years ago.

We can't really fathom time and distance on such a vast scale. We were made to live on this earth, in relation to such objects as trees, cars, desks, mountains, rocks, boats; in relation to prey, predators, and other people. Things we can see and recognize, whether their surface is smooth, rough, soft, hard—or perhaps most importantly, friendly or aggressive. In general, we were made to interact with things close at hand. Not with the universe, not even with the ocean. We imagine the ocean as practically endless, but it is barely a drop in the universe.

Yet we *do* think a lot about the ocean. Maybe, like the universe, our consciousness is expanding.

If you start thinking about the stars in this way, a certain question will usually come floating past: Is there life somewhere out there?

Since there are billions upon billions of planets, and since the universe may well be endless, shouldn't there be a high likelihood that we'll find life? Even if we drop 99.99 percent of the planets because they presumably lack the properties required for sustaining advanced life-forms, there would still be hundreds of billions left.[6] Scientists seem to agree on one thing: Life is most likely dependent on water, no matter where it might be. It's a matter of chemistry. If we assume that the building blocks are the same in the whole universe, then water has to be the essential instigator everywhere, along with carbon. Water doesn't necessarily contain life, but without water, there can be no life. That's why the astrophysicists are not initially looking for life when they study Mars and the other planets. They're looking for water. But mostly they're finding ice and steam, sometimes in amazing amounts. In 2011, two NASA teams discovered a reservoir of steam surrounding a quasar that is twelve billion light-years away from earth. The mass of water was estimated to be 140 *trillion* times bigger than all the water present on earth.

Over the past few years, scientists at Penn State University (at the Center for Exoplanets and Habitable Worlds) have been searching for advanced life in hundreds of thousands of galaxies. They look for unusual quantities of midfrequency infrared rays, based on the theory that highly evolved cultures must use energy that produces heat. So far they haven't discovered anything remarkable.

In the summer of 2015, NASA scientists announced that they had identified a planet outside our solar system that was as similar to earth as any they had ever found.[7] It might be habitable. But its sun emits more energy than ours, so the planet might be simply a desert of rock with an atmosphere, just like our planet will one day become. Today the earth is blessed not only with an atmosphere and enormous amounts of liquid water, but also with nutrient-rich arable soil, which alone can feed many billions of people and animals.

The bar scene in the movie *Star Wars,* in which colorful drinkers from various galaxies meet to fraternize or fight, may be very entertaining. But even if there are many billions of galaxies, the human being may be the only creature in the whole universe to hit a bar. Though I like to doubt that.

Along the deep-sea mountain chains that circle the earth there are many volcanic vents, or "chimneys." In 1977, scientists discovered that they are swarming with life. Gushing out is a sulfurous, boiling fluid, which, because of the pressure, is 750 degrees Fahrenheit. No one thought life could survive under such circumstances. Small creatures do thrive, and some larger species nearby spend their lives in water that is 175 degrees Fahrenheit.

At the greatest depths there is no light, and hence no plants. Energy is created by means of chemical reactions. Toxic substances are broken down by bacteria, to become nutrition for other species. Down there, life is sustained not via photosynthesis but via chemosynthesis. Some scientists suspect that life on earth began around these types of deep-sea vents. Others think life originated way out there in the starry depths.

Hugo takes off his headset, looks around, and brings us out of our trance. I've brought along a bottle of whisky, for special occasions. There's no particular cause for celebration at the moment, but that makes it even more special, so I open the bottle. Hugo doesn't care for hard liquor, but we happen to have our box wine. I take a big swig of whisky. The heat spreads from my stomach like a little Gulf Stream, all the way out to the northernmost and southernmost parts of my body. Our boat isn't drunk, but it reaches a slightly tipsy state before Hugo takes another look around, this time as if he has suddenly realized how late it is. The sea is wine dark, but the stars are shining as if through a perforated lampshade.

Hugo decides to tell me a story. This one is about the time he and his uncle Arne crossed Vestfjorden in the *Helnessund*. Arne was known as a man with a powerful voice. He was good at scolding and ranting, and his bellowings always dominated noisy gatherings, such as celebrating Constitution Day on May 17 or at parties for teenagers in the local village hall. Hugo, who was fourteen at the time, went into the small room behind the boat's wheelhouse where the echo sounder and radio were kept. On the table lay an open notebook along with the sea charts. Uncle Arne had written a poem, and one of the verses has stayed with Hugo ever since: "Beneath the swarm of heavenly stars / I stand here tonight and feel / at my wheel."

Just as the beam of the Skrova lighthouse switches on, Hugo says, "We need to pull up the line and head in. It's almost pitch-dark."

Light shoots through the darkness, capturing us in a blitz-like moment. Then it sweeps onward, sending its rays far out to sea.

Of all the meaningless schemes we could have come up with, this feels meaningless in exactly the right way.

16

A fishing float doesn't get interesting until it moves. And the longer it takes for that to happen, the less compelling it becomes. We go out every day, staying from morning till night. And for every day that passes, the likelihood that something will happen diminishes. Occasionally we pull up the hook to make sure the blubber is still there. No bite marks, only a few little parasitic creatures from the seafloor. Don't Greenland sharks like farmed salmon? Has some other scavenger, like maybe a lamprey, gotten to the bait first? A lamprey can strip clean a huge halibut in a few hours. If it's left too long on the line, the fisherman will pull up nothing but its hollowed-out skin. Or could it be that the whale blubber is so odorless and clean that the shark hasn't noticed it?

We always see the Skrova lighthouse when we're out in the ocean, standing there so stalwart and erect. We pass close by, both on our way out and when we come back in. On the third day I have a feeling that the lighthouse's deranged eye has begun staring at us.

We would have liked to go ashore to see it up close, but because of the currents in the sound, that's easier said than done. It's not so simple to moor such a small boat without hauling it up onto the wharf.

The Skrova lighthouse was built in 1922. During the first decades, two families lived at the station at the same time. That was probably a good idea, since it was well known that lighthouse keepers who spent long periods on their own sometimes lost their minds. Many couldn't cope with the isolation. Perhaps in an attempt to promote mental health, the Norwegian Lighthouse Association had its own mobile book collection that was moved from one lighthouse to another. Some of these books happen to have ended up in my personal library, including a couple of volumes of Icelandic sagas. When I open one of them and see the logo of the Lighthouse Association inside the cover, I think about the book once having made the rounds of all the Norwegian lighthouses, back when they were manned stations. I picture the lighthouse keeper sitting inside, reading sagas from Iceland in the dark of winter, as the storms slammed against the windowpanes and life at the lighthouse was filled with longing and dreams.

Fog must have been an extra burden, because then the lighthouses had to use sirens to warn of their position. From 1959, they used a so-called super-typhoon. It sent out deep, plaintive signals that could reach into your very marrow, even at a distance of several miles.

During the war, the Germans occupied the Skrova station, and a soldier by the name of Kurt supposedly hanged himself inside the lighthouse. That's something people in Skrova have never forgotten, even though it might be nothing more than a myth.

Lately people have been talking about a much more recent tragedy. Not long ago the ferry *Røst* went out into the waters between the lighthouse and Skrova. The crew was trying to measure the distance to the high-voltage power line that stretches across the sound, and someone made a fatal miscalculation. From the top of the ferryboat's mast a seaman tried to measure the approximate height to the line—using a fishing pole. The pole touched the power line, sending twenty thousand volts through the man's body, killing him instantly.

Other countries have magnificent buildings in the form of churches, mosques, palaces, and the like. Skrova lighthouse stands on a little island, just off a somewhat bigger island, out in the ocean. It looks as if it had been airlifted there, all in one piece. Or as if it sprang from the ground on its own, like a plant made of stone, and then grew a little taller each year until it reached the height it was meant to have.

What really happened was far more labor-intensive. The lighthouse itself and the two lighthouse-keeper residences—because they built not one but two big houses on this islet—were brought ashore from boats, stone by stone, plank by plank, in spite of volatile currents and stormy weather. Everything had to be carried by the seamen, construction workers, and engineers.

The reflectors and lenses in the eye of the lighthouse are a happy marriage of several sciences combined. Originally only one demand was made of a lighthouse: it had to be visible from far out at sea, which meant it had to be tall. This functional requirement has created our most harmonious and erect of structures. Their very location—on exposed promontories, cliffs, islets, and small islands at the mouths of fjords—lends the lighthouses an aura of triumph and vigor, as if they were built by a civilization that spreads light in the darkness and is able to defeat the forces of nature. They look best from the sea.

There are two Skrova songs. One is about the lighthouse, as viewed from far out at sea by whoever wrote these words: "Have you ever seen a prouder sight / than Skrova's lighthouse shoreward / gleaming like a bolt of lightning?"

In Scotland, a single family is responsible for all ninety-seven lighthouses that were built along the Scottish coast between 1790 and 1940: the Stevenson family. Robert Louis Stevenson, who wrote *Treasure Island, Strange Case of Dr. Jekyll and Mr. Hyde,* and other classics, was actually supposed to become a lighthouse engineer, in keeping with family tradition. Instead he became rich and world famous as a writer, but he was also considered the black sheep of the family. Unlike nearly all his male relatives—in particular his great-grandfather, father, uncle, and brother—he did not plan, design, or construct lighthouses. The Stevenson lighthouses were often built on reefs that were submerged at high tide, where the North Sea and the Atlantic Ocean collide and create foaming currents and violent swells capable of washing away nearly everything.

For almost seventy years, before the Skrova lighthouse was built on the island of Saltværøya, a so-called fishing lighthouse stood on the islet of Skjåholmen, a little closer to the entrance to Skrova. This old lighthouse on Skjåholmen was the first to be built in northern Norway. The kerosene-burning lamps were lit only from January 1 until April 14, during the wintertime and the Lofoten fishing season.

The Scots have the Stevenson family. Here in Norway we have our suitably modest counterpart in the form of the Mork family from Dalsfjord, in Volda on Sunnmøre. Ole Gammelsen Mork worked on his first lighthouse on Runde in 1825. His son, Martin Mork Løvik (1835–1925) was already the building foreman on Skrova when the old lighthouse was constructed in 1856.

The Mork family produced four generations of lighthouse builders. Unlike the Stevensons in Scotland, they were not innovative architects or engineers. In the summertime, the Morks supervised the work teams that erected lighthouses and navigational markers, or built harbors and roads. In the winter, they fished. The early lighthouses were relatively short structures made of wood and stone, while the later ones were slender, sky-high towers of cast iron. Martin Mork Løvik's son, Ole Martin, built the tallest lighthouse in Norway: the Sletringen lighthouse off the island of Frøya.[8]

The most famous lighthouse keeper at the Skrova station was Elling Carlsen (1819–1900), who in his day was a renowned inventor and arctic sea captain. He grew up going to sea with his father, who was a ship's pilot. At the age of three, and in midwinter, Carlsen was taken along on a trip in a small boat from Tromsø to Trondheim.[9] In 1863, he became the first to sail all the way around Svalbard. Over the next years, Carlsen discovered more islands farther east, in the Kara Sea, where he established good contacts with the nomadic Samoyedic peoples. And in 1871, on the northeast side of Novaya Zemlya, he found the campsite left by the Dutch navigator and arctic explorer Willem Barents, who had discovered Bear Island and Spitsbergen in 1596. Valuable maps, books, and chests filled with equipment, as well as other items, were brought back to Norway and sold to an Englishman for 10,800 Norwegian kroner, which was a huge sum at the time. The following year Carlsen joined the polar expedition of Julius von Payer and Karl Weyprecht as their ice master and harpooner. The goal was to find the Northeast Passage to Asia.

The expedition was sponsored by the dual monarchies of the Austro-Hungarian Empire. During the first winter the *Admiral Tegetthoff* got stuck in the ice. Slowly the ship began to break up, and was twisted into pieces. The crew endured hunger, scurvy, tuberculosis, madness, and infighting; some died. After two winters they gave up hope that the ship would ever come free, and they began dragging three small boats across the ice, in the hope of reaching open waters. Even the levelheaded Carlsen lost all sense of equanimity during this period. After three months of inhuman trials, dragging the boats toward the drift ice, they finally made contact with some Russian salmon fishermen on board a schooner off Novaya Zemlya. The Russians took the desperately exhausted men to the municipality of Vardø, in the extreme northeastern part of Norway.

The expedition is recounted in the historical novel *The Terrors of Ice and Darkness* by the Austrian writer Christoph Ransmayr. For his source material, the author makes use of the Austrian participants' diaries and memoirs. While their ship was frozen in the ice, Julius von Payer went out on dogsled excursions to the north. That was how he discovered Franz Josef Land, an archipelago consisting of 191 islands in the Arctic Ocean, Barents Sea, and Kara Sea. But when he arrived back in Austria, no one was willing to believe him. He painted landscapes of the icy desolation, but his paintings were not popular, and von Payer died destitute and alone in 1915.

As for Elling Carlsen, Ransmayr writes: "The old man, who has spent so many years of his life in the arctic seas, always wears his white periwig when invited to the officers' table; on those feast days of martyrs whom he holds in special honor, he pins his Order of Olaf to his furs. (But when the waves and veils of the northern lights flare up in the sky, Elling Carlsen removes everything metallic from his body, even his belt buckle, in order to prevent any disruption in the harmony of their flowing figures and to ensure that the fire of the lights is not directed toward him.)"[10]

Carlsen was awarded an Austro-Hungarian order for his efforts. In a mini-biography written by one of his contemporaries, the polar historian Gunnar Isachsen, Carlsen is depicted in this way: "In his personal life he was not a happy man, and his two sons met with tragic fates. Those who traveled with Carlsen described him as a skilled seaman and hunter. When he was working on something, he was impossible to satisfy. Otherwise he was a pleasant person; he is even portrayed as unusually amiable."[11]

In 1879, Carlsen was put in charge of the old Skrova lighthouse, and there he stayed for fifteen years. Carlsen must have been quite a tough guy. Yet he was also a vain man and deeply religious, even superstitious. He often sported gold earrings, although maybe not when the northern lights appeared.

At the lighthouse station, whenever the storms raged and he sat amid the kerosene fumes from the lamps, staring out at the sea near the entrance to Skrova, he undoubtedly had time to reflect on his life. He had experienced a great deal and seen lands no one had seen before. For him, the ice and the island realms near the North Pole were not a blank canvas but places teeming with life and marked by distinctive local characteristics, and almost no one on earth knew them better than he did.

It's not Carlsen's old lighthouse that is keeping an eye on Hugo and me. It's the "new" Skrova lighthouse, which was built on Saltværøya in 1922. Like many other lighthouses from that period, the Skrova lighthouse is painted rust red with two broad white stripes. In my mind the structure looks like a slender, stern person wearing a sweater.

The new Skrova lighthouse was designed by Carl Wiig in 1920. He was born in the old Norwegian fishing village of Gjesvær on the island of Magerøy, far up near the Arctic Ocean in Finnmark, just ten miles or so as the crow flies from the

North Cape. His father was a merchant in Leirpollen, a little southwest of the North Cape peninsula. Wiig was only twenty-five and newly hired by the Lighthouse Association when he designed the Skrova tower, although more experienced designers and engineers must have given him guidance and advice. In this instance as well, it was a work team from Volda that did the actual construction. The man in charge was Kristian E. Folkestad.[12] His family, from Folkestad, on the other side of Dalsfjorden, had similarities with the Mork family. They too had built lighthouses along the coast for several generations. In the summertime, nearly every farm on Dalsfjorden sent men north to join the work crews.

The school records from Trondhjem's Technical Institute show that in terms of grades, Wiig was at the very bottom of the class of approximately 250 engineers who took their exams in the period between 1910 and 1915. In other words, an academic loser from Finnmark designed the Skrova lighthouse.[13] I myself come from a place in Finnmark that the indigenous Sámi people call Ákkolagnjárga, which, according to sources, means "Greenland shark promontory." Even scholarly Sámis can't tell me why. As far as I know, the seafaring Sámis didn't go after Greenland sharks. Why should they? This type of shark has inedible flesh, swims at a depth of hundreds of feet, and is impossible to handle from small boats. It just wouldn't make sense.

The eye of the Skrova lighthouse stares down at us as we move past at a speed of six knots, like two microscopic dots in the middle of a churning cosmic maelstrom. Whenever we get too far away and lose sight of the floats, Hugo starts up the motor, and we go back. But most of the time we sit in the boat, only half awake, occasionally chatting or else silently following the calm waves of our own thoughts and associations. Neither of us has begun to question our self-imposed task. On the contrary. We know the Greenland shark is swimming below us, and we're confident that we'll be able to bring it to the surface.

Seals and porpoises stick their heads out of the water. Maybe they're starting to recognize us; maybe they're wondering what we're up to. We belong on land, they belong in the sea. Every time they're in shallow water, every time they look toward shore, they see what for them must be a dangerous and unfamiliar element.

The sea shows us a gray-blue and unusually blank face during these days. The water is smooth and pale, almost lethargic, and the autumn is cool and clean. On both sides of Vestfjorden we can already see snow on the highest peaks. The silhouette of the Lofoten Wall looks as if it has been carved with a sharp knife, but otherwise the slopes are softly contoured, without contrasts or shadows. The sky in the southwest shows delicate threads between clearly defined clouds, reminiscent of marble. "Nothing is as spacious as the Sea, nothing is as patient."[14]

Mostly we chat about what we're experiencing at the moment, but when we're simply waiting and everything is calm, our conversation sometimes switches to bizarre topics. One afternoon I describe how animals from the Middle Ages all the way up to the 1800s could be taken to court for breaking human laws. Dogs, rats, cattle, even millipedes were charged with and jailed for crimes ranging from murder to indecent behavior. Defense attorneys were appointed, witnesses summoned, and every legal procedure of the day was followed. Sparrows were accused of twittering too loudly during a church service. Pigs that had attacked young children were sentenced to death. In France, a pig was dressed in a suit, led to the gallows, and hanged. In 1750, a donkey was found innocent after an unfortunate incident only because a priest was able to testify that the animal had previously led a virtuous life. Today it's difficult for us to understand why people took these sorts of actions. They may have feared chaos and anarchy and believed that nature was also governed by laws of morality.

Hugo asks me whether I've ever heard of Topsy the elephant. I haven't.

"Topsy the elephant killed two animal handlers and was publicly executed in front of a paying audience in an amusement park in New York in 1903," Hugo tells me. Pausing for dramatic effect, he then adds, "They put what looked like copper sandals on the elephant's feet and sent seven thousand volts of alternating current through its body. The original plan was for the elephant to be hanged from a crane, but they ran into some practical problems. The whole spectacle was done for the sake of PR for the park, and the event was filmed by Thomas Edison's film company. The movie is called *Electrocuting an Elephant*."

17

The days of calm, flat water come to an end when a new autumn storm strikes Skrova. Once again we have to take special care to tie up our boat and floating dock securely and then wait until the storm blows over. It arrives from the southwest and sweeps right into the bay. Even the ferries and catamarans stop running. The bad weather keeps me awake at night.

The phantom of the sea howls out in the fjord as he rows his half boat through the darkness of the winter night. Beneath the station the sea slams against the rocks and wharf posts. The wind whistles around all the corners, and the building groans with every stormy barrage. Something—maybe it's the whole roof—is emitting a deep vibration, sounding like a distant chainsaw heard from inside a cabin. Closed sliding doors are rattling on their tracks, and piercing echoes shoot from room to room throughout the station. Both the sea and the wind are blowing through the buildings because there are gaps, cracks, and small openings everywhere, allowing air to be sucked in and creating a draft.

The whole building is filled with sounds, the way a choir or a church organ fills a cathedral. And all the sounds merge in one rich, multivoiced roar. The bright, irregular splashing from under the wharf can be heard above the deep booming from the far side of the bay. The entire station is stretching and creaking like a wooden ship in the process of tearing free from its moorings.

I lie in bed, listening to all of this. Through the roaring outside I notice another sound. It's closer, not as massive or orchestral; it must be coming from inside the building itself. Something or someone is up in the attic, making a sound that reminds me of sobbing. Did a bird somehow get inside? I try to sleep, but the plaintive sobbing doesn't stop. For a moment I can't hear it, and I wonder if I had simply imagined the sound. Then it starts up again. I really should go up there and check, but there's no electricity or lighting in the huge, stumbling space of the attic. And, besides, I'm freezing. I get a sweater from my bag and put it on. I consider going upstairs, but then I climb back into bed and fall asleep.

Restless foam-crested waves wash through me. I dream that I'm standing at the foot of a steep cliff. Before me stretches the sea, swelling rapidly and about to rise up in a tsunami, which pushes in front of it a wall of things that have been brought up from the seafloor: old shipwrecks, dead whales, flotsam. Octopuses shrouded in seaweed and plastic flail their arms like furies. I see big snipefish and bloated, slimy creatures from the dark depths, along with beasts and monsters found only in old books . . . Everything is coming toward me. Since I'm standing on a ledge between the sea and a towering cliff, I can't flee. Then, just as the tsunami is about to reach me, I wake up. Luckily it was only a dream. I suspected as much even while it was going on.

But something isn't right, because again I hear from the attic what sounds like muted sobbing. This time I put on my pants, light a candle, and head upstairs. The draft is so fierce that the candle goes out. I stop midstairs to relight it. As I'm standing there, I clearly hear what I think is the sound of a woman whimpering, and it's coming from the very back of the attic. There are supposed to be only two other people in the building. Hugo and Mette are asleep in their room, which is right next to mine. Neither of them would have considered going up to the attic in the middle of the night. Definitely not, because no one ever uses the attic for anything, especially not as a place to sob.

For long periods we're so isolated at Aasjord Station that it almost feels like being on board a ship far out at sea. If Mette and Hugo were expecting other guests, I would have heard all about it long ago, and under no circumstances would any guest be up in the attic at night. Even an intruder, if such existed on the island, would hardly have found his way up here. The stairs are well hidden in a corner of the second floor in one of the big, pitch-dark buildings. It's true that the doors are not usually locked, but if someone crept inside to seek shelter, they'd have about thirty rooms in which to hide. They wouldn't have found the attic even if they tried.

It has to be an injured bird. Or maybe an otter? No, an otter would sneak in and steal dried cod from the ground floor so it could dive into the sea if anyone came. There's no way an otter would climb upstairs. A stoat? Too risky for that type of animal too, venturing through room after room, from one floor to the next, with fewer and fewer options for retreat. Well, maybe it's a bird. But that's not what it sounds like. A bird would be beating its wings and not whimpering like an unhappy woman.

The first thing I notice is that the attic floor is slippery and wet, as if something slimy has dragged itself across the floorboards. The sound is getting clearer, and I'm absolutely positive it must be coming from a child or a woman. I decide it has to be a woman and now it sounds almost seductive. A melancholy humming carried in from the sea and almost drowned out by the wind. But it's not the wind or the sea that is singing. There are no walls between me and the voice as I move farther into the attic. With only the faint glow of the candle, I have to be careful not to stumble on a seine or cut myself on a rusty barrel hoop.

The seductive song of the Sirens caused seafarers to run aground. Circe transformed Odysseus's crew into swine. In the corner of the attic I can now make out a shape. I'm not scared, because something tells me that whatever it is over there, it won't or can't hurt me. The silhouette is hazy. I approach cautiously as I try to figure out what the shape in front of me could be. I see long, fair hair, a naked torso with breasts, but the lower body . . . it's a fishtail, it's a . . .

Then I wake up, bathed in sweat, as if I've been in the ocean.

The next morning I open my eyes, feeling as if I've been ill. Hugo says he heard me shout through the wall of my room in the night. I tell him I was about to drown in my own dark sea swells. Then he says that he also heard me get up and walk around. I tell him I don't remember that at all.

18

Since we can't go out to the fishing banks, the Greenland shark is living large in Vestfjorden, free of the threat that so often hovers above it in the form of two motivated men in a tiny RIB.

On the second stormy day (though the wind seems to have decreased to a stiff gale) I take a walk along the cliffs and shoreline on Skrova. The sea is gunmetal gray with big white breakers. During the night the water has been churned up and swirled around. On the headlands facing the sea, I come across lots of pollock that have been flung ashore by the storm. The currents must have whirled the fish up through the water column to the surface and then cast them onto the beach. They can't have been lying there for long, since they would have been eaten by the wild otters, mink, fox, crows, gulls, or sea eagles. A short distance away I find a dead seal already bloated with gases.

On the Orkney Islands they have legends about *selkies,* "seal men," who can swim like seals in the ocean. On land they take the form of ordinary men, except that they have unusually gentle features—something that makes them especially dangerous to young women. In northern Norway, people used to fear the *draugen,* the ghost of the sea, which was ascribed very different traits. The *draugen* was supposedly the phantom of a drowned fisherman, with dead red eyes and wearing an old-fashioned vest made of leather. His head was nothing more than a clump of seaweed. Whenever he appeared, in half a boat with tattered sails, he liked to pull alongside the boats of the living. If he screamed and carried on, it was crucial not to reply. The *draugen* was a portent of death for everyone who saw him, provided he didn't take them down into the sea then and there. He could spell death even without being seen. In the night he would mess with the equipment on board while a boat was docked. If the oars were turned with the blades facing forward, things did not bode well for those who sat in the front of the boat.[15]

Throughout his life, Hugo has known many older fishermen who have very definite ideas about the *draugen.* They don't consider the stories about him to be just folklore or mythology, but something much more real. If you asked them directly, they wouldn't admit to believing in the *draugen,* because they know that would make them sound foolish. But this malevolent spirit of the sea still lingers in the mind of some fishermen.

At the end of the beach I have to climb over a pile of rocks. On the other side I come to a new stretch of shoreline. It has been completely washed clean, without a single trace of kelp or greenery. The sea has taken everything. At one end is an old boat launch with rusty rails that go down into the water. When

I was a kid, I often saw railroad tracks laid down on sloping rocks and shores. They were used to move boats up or down from a boathouse or slip. But in my imagination they were built for trains going down to the ocean bottom, with watertight compartments from which fantastical sights could be seen.

I continue on past the rocks as the storm rages, getting stronger the farther west I go on the island. The sky rushes blue black and low over the sea and islets. Cymbals and bass drums crash simultaneously. I was once in a hurricane, and what I'll never forget is the sound. Ordinary storms whistle and howl. In a hurricane all these high-pitched, familiar sounds seem to disappear. What's left is a deep, dark, penetrating sound, as if the soul of the universe were making itself known with a cold fury.

There's a salty, fresh, but slightly rotten smell in the air at the moment, like when the bodies of two people have joined on a hot and humid night in a bedroom with the windows closed. The water is rushing in through small, narrow cracks in the rocks, shooting up like geysers when it reaches the end and collides with the mountain. Each time the water takes several tiny pieces of stone from the cliff, carrying them back out to sea. Maybe one day they'll form a new beach along some faraway coast.

From the crests of the waves the wind grabs tiny drops, which then sweep like weightless rain toward land. When the waves slam against the rocks, they shatter and turn to spray. Water molecules dance around on the world's oceans, dissolving, evaporating, cooling, and combining in new ways. The drops that strike my face have been in the Gulf of Mexico, in the Bay of Biscay, through the Bering Strait, and around the Cape of Good

Hope many times. Maybe over the eons they've actually been in all the oceans, both big and small. In the form of rain they have washed over dry land; there they have been lapped up thousands of times by animals, people, and plants, only to evaporate, transpire, or run back out to sea, again and again. Over billions of years the water molecules have been everywhere on earth.

The sea slams against the cliffs and rocks with thundering crashes and sharp, hissing sounds. The wind dissipates the clouds, but the sun never makes an appearance. The horizon is saturated, and the light seems to be coming from the gray-green water, which is hammering the shore. I'm suddenly scared that the sea might reach to where I'm standing. No, that's not right. I'm gripped with an irrational fear that the sea *is trying to do that.* Even though I laugh out loud at such a foolish idea, I climb up onto a bigger rock. Even the seagulls have flown to higher ground to hide.

The sea is the first source. Waves from a deep, primordial past flow through us, like echoes of small splashes inside an inaccessible cave near the ocean. Sometimes, when we stand on shore during a powerful storm, it's like the sea wants to take us back. A wave far out on the horizon slowly starts to build up extra muscles—you'd think it knew beforehand exactly where it wants to go and how it's going to get there. The wind helps out; the movement and rhythm are perfect all the way to shore. Other waves give the first wave a shove, cheering it on and allowing it free passage. As it approaches land, it picks up speed and gathers its forces to leap.

On shore, let's say that a couple, newly in love, are taking a walk. Or maybe we see a morose couple from the Czech Republic, a local amateur photographer with a new camera, or a bunch of inquisitive teenagers who are bored with being at home and who still haven't discovered that they could die. All of these people left their safe, warm houses, their comfortable cabins and hotel rooms, to come out and feel the ferocity of the storm on their own bodies.

They walk along, shivering a bit but mostly enjoying the forces at play outdoors, though from a safe distance. Maybe looking at the stormy sea makes one of them realize how old the earth is. They take note of the whole vast surface in which the wind is creating furrows on the wave front, the foam whipped up like white hair, the low rumbling sound, and everything else that gives the sea its prehistoric face.

In the old days, people called big waves *brimhester* (breaker horses) because the crests as they raced to the shore reminded them of the manes of horses. Now the spectators see a huge wave—they didn't realize the sea was even capable of producing such a thing. It's headed for land, arching its back as it opens its mouth wide. A tongue of searching seawater stretches out, higher and higher, much longer than any of the other waves, far past the foreshore and over the steep slopes. This wave is like no other. It pushes past the barriers of boulders, rocks, and crags, then continues for many yards up onto what otherwise belongs only to land. Like the arm of an octopus, the wave shoots out from the sea, aiming for the spot where someone is standing, completely unaware of what is about to happen.

The current is so strong that it sweeps the people off their feet, even though it doesn't even reach to their knees. If not for what happens at that brief moment, they could have stood in that very spot for the next fifty years or so without ever getting their feet wet. They came out here on a whim when they could just as well have stayed home and done what people do when they're living their normal lives from one day to the next.

The wave topples them over. This could still have been just an amusing story to be told at lunch the following day if not for the fact that the water needs to go back out to sea, and takes along everything in its path. Hands fumble desperately for something to cling to as they lose their grip on the slippery boulders and slick rocks or are scraped bloody by barnacles. Kelp and sand fill the hands, but to no avail. The undertow is too great. First bewilderment, their eyes meeting for a tenth of a second, an inquiring look, wondering whether the other person has somehow devised this prank. Then both people realize it's serious. Shock and panic race like white bolts of electricity through their bodies. Their brains experience this moment in several dimensions at once. Time freezes. Adrenaline pumps, and all their alarm systems go off. What was supposed to be a nice outing along the shore in rough weather, maybe in order to build up an appetite before dinner, is about to become a brief last act. The curtain rises, life unfolds on stage, not as a farce but as a revue in color and with the fast-forward button pressed down.

The sea sucks in its tongue, licks its lips, and closes its mouth. Only a few short-lived streaks of white foam are left on shore. In the ocean the people tumble around as if inside a washing machine, until they can't tell up from down. Maybe they get slammed against the rocks and are already unconscious before they drown, as they're dragged farther out, farther down. Maybe their bodies will never be found. Gone. Forever. These kinds of accidents happen every winter and fall, when storms rage along the coast.

Toward evening, the sea thunders from the west. Big black patches of cloud glide in over Skrova to cover the moon. When the power goes out, it's pitch-black. A bottomless night comes racing in with the storm, forcing its way into everything and everyone.

19

By morning the storm has calmed a bit, but our little floating island is still in the midst of a churning sea. According to the weather forecast, it will take days before we can even think about venturing out on the water. So I stay inside, reading and taking notes, while Hugo continues with his carpentry work on the Red House. Fortunately, he's far enough along that he can work indoors as he listens to the radio. He's wearing his headset, which he has a tendency to take off and leave in the strangest places. He's constantly trying to remember where he put it.

The bad weather gives me the opportunity to read the books I've brought along. I get out a massive volume with a white cover, which was first published in Latin almost five hundred years ago. I know that the author, Olaus Magnus, wrote about the exquisite monsters that, in his day, were found in the oceans, especially off the coasts of Norway and Iceland. It so happens that he drew the sea monsters he describes in his book on a map, and I'm familiar with that map: Olaus Magnus's *Carta marina* from 1539.

Olaus Magnus was the Latin name of the Swede Olaf Måns-son, from the town of Linköping. He was a Catholic bishop, but he was forced into exile—first in Gdansk, then in Rome—when Sweden became Lutheran. In Rome, he worked on *Carta marina* and his epic history of the Nordic peoples, which was published there in 1555, under the patronage of Pope Julius III. The history is divided into twenty-two books and 778 chapters. In my Swedish edition it comprises more than eleven hundred closely typeset pages, all in one volume. The book proves to be a sumptuous treasure trove. Olaus Magnus was an eminent humanist, in the sense that he was extremely learned and sought knowledge from all kinds of sources, turning in particular to the classics of antiquity.

In keeping with the tradition of his day, the full title of his book says a lot about the nature of the contents: *A Description of the Northern Peoples, Their Various Relationships and Circumstances, Their Habits, Their Religious Practices and Superstitions, Their Skills and Occupations, Societal Customs and Ways of Life, Their Wars, Buildings and Tools, Mines and Quarries, and Wondrous Things About Nearly All the Animals Living in the North and Their Natures: A Work of Diverse Contents, Replete with Far-Flung Knowledge and Partially Illuminated with Foreign Examples, Partially with Portrayals of Domestic Things, Intended to a High Degree as Amusement and Entertainment, and Meant to Leave the Reader in Delighted Mind.*[16]

Olaus Magnus wrote an important work, which, over the

course of the next centuries, was translated into English, German, Dutch, and Italian. His goal was to collect everything he could possibly discover pertaining to the north. By the second book section, he already begins describing "the great quantities of wondrous phenomena that belong to the water's element, especially in the endless ocean that touches the northern part of Norway and the area's countless islands." In this section the author discusses, for example, Iceland's volcanoes, where for him the spirits and shadows linger from those who have drowned or suffered particularly violent deaths. Their ghostly figures can assume forms so they can't be distinguished from the living, though they avoid shaking hands if you meet them. Among other things, Olaus Magnus also describes the horrifying sounds issuing from some beach grottoes, the stench of dried fish, the strange nature of ice, the kayaks of the Greenland Inuit, the mysterious cliffs of the Faeroe Islands, the fathomless deep off the Norwegian coastline, and the rivers of northern Sweden.

Parts of Olaus Magnus's work function as a supplementary description of all the fantastical creatures depicted in detail on his famous *Carta marina*—in particular the monsters. Today the map is celebrated, but sometime around 1580 all known copies disappeared. Not until 1886 was a copy discovered in the national library in Munich, Germany. And in 1962 another copy was found in Switzerland and acquired by the university library in Uppsala, Sweden. It's painful to think that this treasure might have been lost forever.

Olaus Magnus had done extensive research while making lengthy trips on both land and sea, within Scandinavia and beyond. It's uncertain whether he personally visited the coasts of Norway, though he writes so much about them. But his work is encyclopedic, and many descriptions are undoubtedly based on the stories of fishermen and seafarers. And, to an even greater extent, on everything that was written by antiquity's numerous known and unknown authorities regarding various phenomena and forms of life in the sea, which possesses "a heavenly and eternal fecundity."

Like other learned men of that time, Olaus Magnus believed that all animals found on land have their counterparts in the sea. The same is true of plants. For all those found on land, including coltsfoot, there is a similar variety in the sea. According to Olaus Magnus, the ocean also has its own variations of birds, plants, and animals—everything from lions to eagles, as well as swine, trees, wolves, grasshoppers, dogs, swallows, and so on. A very long list. Some animals grow big and fat by breathing in the south wind; others by inhaling the wind from the north.

In addition, there are many mysterious animals that belong only in the ocean and nowhere else.

Included on the *Carta marina* are drawings of forests, mountains, towns, people, and animals on the Scandinavian peninsula, as well as on the landmasses of Denmark, Scotland, the Faeroe Islands, the Orkney and Shetland Islands, and Iceland. On the Orkneys, the animal life on land has been given a fairy-tale quality. It was said to include a tree that bears fruit from which live ducklings hatched.[17]

But it was the extremely lifelike drawings of monsters in the waters between these countries that would make Olaus Magnus's *Carta marina* famous. The northern and western sections of the map, which are mostly ocean, are illustrated with beasts, one more sensational than the next. Some have hellish, glowing red eyes and fangs in their lower jaw. Others are capable of swallowing ships whole and are filled with evil intentions. Or their size alone may make them deadly dangerous. In his book, Olaus Magnus provides detailed descriptions of how unaware seamen might drop anchor on the back of a sea monster and then light fires to cook their food, believing they were on dry land. Of course, the heat from the fire would wake the gigantic fish and make it dive. The men on its back would be dragged all the way down into the deep.

Carta marina depicts sea unicorns, giant flying fish, sea cows with horns, sea rhinos, sea horses as big as bulls, poisonous sea hares, sea mice, and a polypus, which with one of its ten claws could lift a man out of a boat and carry him down to its family, waiting on the seafloor.

Life could not have been easy for seamen in the days of Olaus Magnus. And those who happened to see copies of his map, or maybe were even educated enough to know Latin and hence were able to read his book, must have been terrified. They were already familiar with many of the sea's dangers. But the learned Olaus Magnus's amazing catalogue of horrifying monsters surpassed any rumors they might hear in even the worst flea-ridden tavern down at the harbor. It would have been perfectly understandable if a man started looking for work on shore.

What are you supposed to do, for example, if you encounter a ziphius, as depicted in the sea off the Faeroe Islands? This gigantic monster has an owl's face with a nasty curved beak. It uses its dorsal fin to jab or saw huge holes in the bottom of ships, and through the holes it then devours the crew.

And what about the hairy sea swine? It looks like an enlarged pig, but it has four dragon feet and two eyes on either side of its body, as well as an eye on its belly, near the navel. Sometimes the sea swine is accompanied by its "companion," the sea calf. Each is bad enough on its own, but if they appear together, they incite each other to new heights of malice. These two assailants are among the worst you could ever meet.

Olaus Magnus writes that a sea swine was observed in the "German ocean" in 1537. This prompted the Vatican to initiate a study into what its appearance might augur. The learned men in Rome decided that the presence of the sea swine was by no means a good portent, and the papal commission finally concluded that the animal symbolized a distortion of the truth—not the *stories* about the animal, but the beast itself, with all its perverse physical attributes.

The book also contains a great deal of practical advice for seafarers. Olaus Magnus writes, for instance, that some monsters will stay away if you blow war trumpets. This works for the spouter, or sperm whale (*Physeter*), which can spout huge gushes of water and, in the worst case scenario, sink even the strongest of vessels. The whale finds the sound of war trumpets so torturous that it will "flee" back to the fathomless deep. As his source for this fact, Olaus Magnus cites ancient Greek and Roman authorities on geography and natural history, such as Strabo and Pliny the Elder. Seafarers are also advised to hurl big vats and barrels after the sea monsters, because that might make them start playing instead of attacking. But if that doesn't work, you could always shoot at them with catapults or cannons; the loud explosions might chase the monsters away.

Ships could also be attacked by birds, a type of quail that settles on masts and sails in such numbers that even the mightiest of ships will sink. In this case, the crew should light torches. And by the way, not all dangerous fish are enormous. There's a fish that is only six inches long. In Greek it's called *echeneis,* and in Latin *remora.* Locals call it the "ship holder." As the name clearly indicates, these fish cling to the ship and hold it in place. The winds may blow, storms may rage, but the ship cannot budge. It's as if the vessel is rooted to the spot. Olaus Magnus received this information from Isidore of Seville (ca. 560–636). But Aurelius Ambrosius of Milan, better known in English as Saint Ambrose (ca. 340–397), also mentions the phenomenon and calls the ship holder a "bad and pitiful sea animal."[18]

Olaus Magnus was an extremely learned man. His work presents a panoramic picture of customs and phenomena in the north, with many described in precise detail. But he doesn't think the way we do today; he divides up the world very differently. The title of chapter 8 in book 21, for instance, is "About hostility and harmony among certain fishes." Here, as in many other places, Olaus Magnus reveals that he assumes fish not only possess a consciousness but also free will, morality, and culture. Some fish live in a natural state of harmony, like the baleen whales. Others are hypersocial and live in huge shoals. But even the herring and other shoaling fish have one particular individual that leads the way, just like humans.

Olaus Magnus writes that in the fish world there are also loners. For some of them, it's simply "impossible to have any comrades," since they spend their whole lives in a state of hos-

tility toward others. The Greenland shark definitely belongs to
this category.

Olaus Magnus had read all of antiquity's known authorities, and he doesn't hesitate to quote them. In the opinion of
Saint Ambrose, for instance, all animals, both on land and in
the sea, possess at least one positive trait that humans would
benefit from emulating. In several places in the book, such as
the chapter titled "A beautiful comparison of fish and people,"[19]
Olaus Magnus describes the enormous parental love that some
fish feel. A callous desire for gain is largely unknown among
fish, since they have no interest whatsoever in material goods or
money. Doesn't the story about Jonah in the belly of the whale
demonstrate that piety is greatest in the sea? The people had cast
Jonah out, but the fish welcomed him. Olaus Magnus's readers
were aware that the story about Jonah was supposed to point to
the death and resurrection of Christ. As Olaus Magnus writes,
Jesus saved not only the earth, but also the sea.

In this epic work, no area is described in a more dramatic manner than the waters where Hugo and I are spending our time
drifting and bobbing in a rubber boat, trying to catch a Greenland shark. Olaus Magnus writes, "Along Norway's coasts, or in
the surrounding sea, wondrous fish abound that have no name.
Yet they are thought to be whales. Their wildness is apparent
at first sight, for they induce fear and terror in everyone who
encounters them. They are gruesome in appearance with their
blocky heads, which are covered with spikes and points and
encircled by long horns that resemble the roots of a toppled
tree . . . When it's dark, the fisherman can observe from a great

distance the glowing eyes, like blazing fire, among the waves."[20] The creature also has hair that looks like goose quills that are thick and long and might also remind someone of a dangling beard. Compared to the rest of the body, the head is small, writes Olaus Magnus. Yet this creature is able, with the greatest of ease, to overturn and sink huge ships filled with strong seamen.

The part of Olaus Magnus's amazing book that is of special interest for Hugo and me deals with sharks, or sea dogs, as some people call them—though not in Norway, where they're called *håfisk* (the Norwegian name for the Greenland shark is *håkjerring*). In the chapter titled "On the gruesomeness of some fish, and the kindness of others,"[21] Olaus Magnus comments directly on a scene depicted on his map. The illustration shows a man being attacked by sharks in the sea southwest of Stavanger. But one of the nice fish, more specifically a skate, comes to the man's rescue. Olaus Magnus explains that sharks will attack in large groups and exhibit unusual ferocity. Making use of their weight, they're able to drag people down into the deep, where they consume their softer body parts. But a skate intervenes to put a stop to this "abuse." The skate attacks with anger and then protects the man until he swims away or, if he's dead, until his body floats up when the sea "cleanses itself."

The sharks, which are innately malevolent, lurk beneath vessels, waiting to grab people. They attack noses, toes, fingers, and genitals, especially keen for every pale part of the human body. Could this be one of the first, highly unreliable, descriptions of sharks attacking people? Could it offer tenuous support for Hugo's speculations on the same topic?

—

Olaus Magnus refers to the learned Albertus Magnus, also known as Albert the Great (ca. 1195–1280), who claims that dolphins will always carry drowned or drowning people to shore—except for those who may have eaten dolphin flesh in their lifetime. As early as the fifth century BC, the Greek historian Herodotus tells of a poet and musician named Arion who is thrown overboard from the ship that is supposed to take him home because the other seamen want to steal the prize money he has won. Arion is allowed one final wish: to sing a song. And with this song, he summons the dolphins, who carry him safely to shore.

Maybe Olaus Magnus had seen in Italy the famous marble sculpture *Boy on a Dolphin,* created by Lorenzetto, a contemporary of his. (Today the sculpture is in the Hermitage in St. Petersburg.) It shows a naked boy, his arms flung out, sleeping on the back of a swimming dolphin. The dolphin, which is only slightly bigger than the boy, displays a determined expression. It knows—just as we, the observers, know—that it represents goodness, and it must rescue the vulnerable human child.

Olaus Magnus explains that both old and new monsters are most often discovered off the coast of Norway because of the sea's unfathomable depths in those areas. In spite of all the dangers surrounding them, the fishermen of northern Norway dare to venture far out in the ocean, where they are constantly encountering the most fearsome beasts.

Not far from the area that Hugo and I have chosen for our hunt, directly south of Lofoten, is the location of perhaps the most spectacular creature of them all. It's a huge, bright red sea serpent, at least two hundred feet long. On the map it's coiled around a big sailing ship, holding a man in its maw.[22]

—

Over the next centuries, Olaus Magnus's descriptions of this monster were to spread far and wide. This becomes eminently clear when you read the book *Norges naturhistorie* (The Natural History of Norway) from 1752, written by Bishop Erik Pontoppidan of Bergen. He describes and discusses a number of *monstris marinis.* The proof of their existence was overwhelming, with special emphasis on eyewitness accounts, many of which were from northern Norwegian fishermen.[23] Pontoppidan can't help but conclude that the monsters do exist—just like the huge snakes in Ethiopia and other African countries. Snakes that, according to reports, are big enough to devour elephants after twining around their legs and toppling them over.

Olaus Magnus also writes about the kraken, the mythological giant octopus, which was thought to live off the coast of Norway. The Icelanders called it the *hafgufa.* As a reliable witness who could vouch for this creature's existence, Olaus Magnus cites Erik Valkendorf, the archbishop of Nidaros (today the city of Trondheim), who, in "the year of Our Lord 1520," wrote to Pope Leo X about the monster. Two hundred years later, Bishop Pontoppidan's descriptions are equally extravagant. He claims a kraken exists that can grow as long as an English mile. It has horns as big as a ship's masts, and it lures fish into its mouth by means of its special smell. When it dives, it creates a tremendous downward suction. The kraken, also called the "Crab" or the "Harrow," is "without doubt" the largest sea monster in the world, according to Pontoppidan.[24]

In the Middle Ages, people thought there were mermaids and mermen off the coast of Greenland.[25] Five hundred years later, they seem to have moved closer to Norway, if we're to believe Pontoppidan. The bishop of Bergen cites several reliable accounts from witnesses who had seen such creatures in the ocean off Denmark and Norway.

One such report came from a Danish mayor by the name of Andreas Bussæus after three ferryboat men claimed to have seen a merman. This caused enough of a stir to launch an official investigation. The merman looked to be of an advanced age but broad-shouldered and very strong. His head was small, his eyes deep set, while his hair was curly and reached only to his ears. He had a lean look, with sharp facial features and a short beard that had apparently been trimmed. From his waist down, his body was as sleek as a fish. Twenty years earlier, the witness Peter Gunnersen had seen a mermaid with long, flowing hair. But perhaps of even greater significance, she had strikingly large breasts.[26]

In Lofoten of old, there were also reports of mermen similar to those mentioned by Olaus Magnus and Pontoppidan. They had the torso of a human and the lower body of a fish. These *marmæler* (merpeople), as they were called, were generally much smaller than the *draugen*. In fact, the smallest of them measured no more than a half inch.[27]

Pontoppidan's book of natural history is illustrated with technically advanced and hyperrealistic copper engravings of Norwegian animals, birds, insects, and fish. Two of the pictures show a big sea serpent in the process of sinking a ship. Pontoppidan's work is worthy of the Age of Enlightenment, since he was a stringent rationalist who wanted to separate superstition, myths, and fairy tales from hard facts. On one hand, no one can refute that the sea is filled with unusual and wondrous things. On the other hand, Pontoppidan has no desire to seem naïve. The sea captains and fishermen, who are the source of such stories, may have misinterpreted or exaggerated what they saw, or their testimonies may have been distorted by others.

For instance, Pontoppidan doesn't believe the Old Norse saga about the merman who was allegedly held captive for a week by some fishermen in Hordaland, on the western coast of Norway, and who sang an unpleasant ballad for King Hjorleif. Or the story about the mermaid who called herself Isbrandt and supposedly conducted long conversations with a constantly drunk farmer on the Danish island of Samsø. But even though Pontoppidan finds the reports of mermen and mermaids to be exaggerated and overly embellished, he does believe in their existence—just as he believes in sea horses, sea cows, sea wolves, sea swine, seal dogs, and the other creatures that Olaus Magnus also describes.

Carta marina is most likely an attempt to depict reality as viewed by Olaus Magnus and his informants—from antiquity's writers to northern Norwegian fishermen. Though it can't be ruled out that the Norwegian fishermen may have been poking fun at the learned bishop who came and asked them about all sorts of things, dragging along a less-than-competent interpreter. It's possible that some of them deliberately exaggerated their stories. But not in every instance.

Great mapmakers like the German cartographer Sebastian Münster (1488–1552) and Abraham Ortelius (1527–1598) of Belgium were nearly as generous with their illustrations of monsters in the oceans of the world. Even Hans Egede (1686–1758), known as the apostle of Greenland, recorded eyewitness accounts of monsters that are no less fascinating than the drawings on *Carta marina*. By the way, from 1707 to 1718, Egede was pastor of the Vågan municipality in Norway, which includes Skrova.

In 1892, when the Dutch zoologist and insect expert Antoon C. Oudemans published a critical monograph on the great Norwegian sea serpent, he was able to list more than three hundred written sources in which the monster was mentioned. Olaus Magnus had set the whole thing in motion, but belief in the sea serpent was still thriving at the end of the 1800s. In his extremely exhaustive book, Oudemans clearly reveals many eyewitness accounts to be no more than lies and deception. The Dutchman's work became the scientific death knell for the existence of the monster. But even Oudemans did not completely dismiss the fantastic; he made his own contribution in crypto-zoology. In his opinion, many observers had actually confused the sea serpent with a gigantic sea lion–like creature (*Megophias megophias*). Yet it too was a mythical beast.[28]

Olaus Magnus, Erik Pontoppidan, and Hans Egede all lived in a time when very little was known about whales and other animals and fish in the deep—and before modern science had established principles for classifying life on earth. Many of the life-forms actually known to science today are even more unlikely than those Olaus Magnus depicted.

Somewhere in his book Olaus Magnus mentions "polyps," described as "many-footed creatures." They have eight feet with suction cups, and four of the feet are extra long. (The octopus has, of course, two tentacles that are much longer than its other six arms.) On their backs these polyps have "pipes" through which the water runs in and out. They have no blood, and they live in holes on the seafloor, changing color according to their surroundings.[29]

If we compare this to what scientists today actually know about squids and octopuses, this is a realistic description. For instance, what does the vampire squid (*Vampyroteuthis infernalis,* or the "squid from hell") do if it's attacked way down in the deep? In the pitch-dark, it wouldn't do much good to spray ink to fend off the attacker, but the vampire squid has another means of defense. It bites off one of its eight arms, which then drifts through the water alone, with tiny blinking blue lights. This diverts the attacker and gives the vampire squid an opportunity to escape. The vampire squid, which can live at a depth of five thousand feet, earned its name because of its eyes, which are the biggest in the animal kingdom, when compared to its body weight. Usually the eyes are light blue, but in a fraction of a second the squid can turn them as red as blood. It looks like a special effect in a cheap horror film.

Olaus Magnus wrote that the voracious *håfisk,* or shark, will even eat small pieces of itself, if necessary. Some types of cephalopods, when in need of food, will actually eat one of their own arms. It grows back. Perhaps even more impressive is the fact that many squids and octopuses are capable of issuing clouds of ink that take on the same shape as their own body—and in some cases, the ink glows with luminous particles. Humans with the same capabilities are familiar to us. They exist in comic books and films and are called superheroes.

A cephalopod discovered in Indonesia in 2005 is able to assume the same shape as a flounder, a sea snake, or almost anything that appears before it. Most squids and octopuses can also instantly change the color and pattern on their own skin in order to blend in with their surroundings. Those that swim in the deep waters are invisible from both above and below.

Their arms or tentacles can shoot out like projectiles, moving faster than our eyes can see. Each arm has a long, heavy suction cup, which in turn has chemical receptors that function like taste buds, while a fine-meshed net of nerve fibers makes the arm extremely sensitive.

Some cephalopods can swim at speeds up to twenty-five miles an hour. They have blue blood, three hearts, a brain in each arm, and nerve cells like ours, but we don't know whether they ever sleep. There is no doubt they are intelligent, and they quickly learn to recognize symbols.[30] And it's clear that they can grow to be enormous.

Even today, only two intact examples of the colossal squid (*Mesonychoteuthis hamiltoni*), which is the heaviest in the world, have ever been studied. Colossal squids live only in the vast depths of the Antarctic and nearby areas, but we hardly know more about them than people did in Olaus Magnus's day. He tended to exaggerate when it came to the size and aggressiveness of sea monsters. Cephalopods do not get as big as ships. But in reality they have far more amazing characteristics than Olaus Magnus described. And it has actually been well documented that dolphins have indeed saved people from drowning.

20

After four days the weather eases. I put down my books and leave my improvised study chamber at Aasjord Station. The storm has left behind a faded, damp world in almost transparent tones of gray. The landscape and the buildings seem to have lost all contour, while the sea is heavy and listless, as if exhausted after the ravages of the past few days. Even the fish I observe from the dock are swimming sluggishly, maybe for lack of anything better to do.

Behind a gray and despondent haze, the sea is pushing in and out of Vestfjorden. The tidal water picks up waves from the south to the north, reaching the Skrova lighthouse twice a day when the sea rises on its coastal currents that slip into the fjords while the formidable currents in the Atlantic Ocean continue on their way toward the Arctic Sea. We could have gone out in the boat now, but I have to return to Oslo.

Hugo and I have already started planning our wintertime fishing trip. We have to get hold of some stillborn or deformed piglets from one of the pig farms in Steigen. Then we're going to catch a Greenland shark. That's what we tell each other when we say good-bye. Do I glimpse a shadow of uncertainty in his eyes?

No, I'm probably just imagining that. We still feel the drag of the irresistible arm Melville wrote about. Our internal mine workers will tirelessly persevere. Our lack of success so far only serves to harden our steely determination. It's like a propeller that spins and spins, issuing a deep, rotating sound. Two men in a small boat, never sure what they might encounter out on the sea or what they might pull up from the abyss, beneath melted stars and electric full moons, where breakers and swells assault the islets like hysterical herds of cattle and the lunatic eye of the lighthouse never lets us out of its sight.

Winter

21

It's early March the next time I travel north, enticed as usual by the sea's old promises of adventure and the hunt for a shark you can only dream about when inland. My route takes me from Berlin to Oslo to Bodø, and then by catamaran ferry north to Skrova. In the villages of Brennsund and Helnessund, white smoke rises from the chimneys through the frosty arctic air.

It's unusually cold. Winter along the coast is often damp and raw, but rarely does it get this cold. Each day the Gulf Stream carries as much heat to Europe as is generated by the entire world's coal consumption over the course of ten years. Lofoten lies a good deal farther north than Nuuk, the capital of Greenland, yet in Lofoten the average temperature is almost ten degrees warmer, year-round. Without the Gulf Stream, the Norwegian coast would be one continuous expanse of ice, broken only by short, arctic summers.

On the catamaran I buy the local newspaper. I read a story about more than a hundred wild sheep that have been taken by high tides. They were on the foreshore in freezing cold weather, near Burøya. Their wool became coated with ice, the tide came in, and the rocks ended up submerged, which was something the sheep couldn't have foreseen. They didn't have a chance. One hundred and four sheep are no longer among us. Only three survived. What were they doing out there on the rocks?

Hugo has had a rough night. Yesterday he treated a floor with lye. But it's five degrees Fahrenheit, which is abnormally cold for Skrova, and the water froze in the pipes. To rinse the lye off the floor, he had to bring salt water from the sea. The result is visible on his fingernails, which have split and partially dissolved. Plus he has the flu.

Yet Hugo is his usual positive self. I ask for an update on the restoration process. He tells me there has been some trouble with the financing needed to realize his big plans to bring a gallery, restaurant, pub, and overnight accommodations to Aasjord Station. The work has come to a standstill, but this doesn't seem to be worrying him too much. I ask about his stomach. He merely rolls his eyes and takes me around the fishing station to show me what he and Mette have done since my last visit.

He's made good progress on the interior of the Red House. They've also done a lot inside the station itself. What first meets the eye is that they've actually managed to clear out the fish-drying warehouse, the storage area for fishing gear, and the ground floor of the main building. All the old seines, vats, tools, materials, and equipment are now gone. In their place Hugo has built a bar from fishing crates printed with the names of fish landing centers and other companies from Vardø to Ålesund. In a week there's going to be a big party at Aasjord Station.

During the winter Hugo and Mette have also spent long periods in Steigen, where Hugo completed several large oil paintings. One painting is twenty-three feet wide and it is going to be hung in Stormen, the magnificent new cultural center and library of Bodø. A couple of other unfinished paintings are of the Greenland shark. The huge elusive fish has begun to impact our lives, maybe even to the point of turning Hugo into a figurative painter.

As for storms: Hurricane Ole swept across Skrova and blew

a large shed off a dock and into the sea. It also sent the waves surging up past Aasjord Station. The sea reached so high, they actually had to remove some floorboards inside the station to let the water flow *in*. Otherwise the waves would have forced their way through the floor and caused much more damage, maybe even lifting up the whole building. If the station hadn't been restored and was still perched on wobbly rotten posts, it probably wouldn't have withstood such extreme weather.

After Hugo has brought me up to date, I ask him, "So, what else? What have you been doing with your time?"

"What else?!"

This time I can't keep a straight face and start to laugh.

"What about you?" he asks.

"Oh, you know. Dealing with all the big-city hustle and bustle. Nothing but dirty snow, caffé lattes, fish sticks, bad kebabs, parking tickets, and stress," I tell him.

Hugo laughs. He has nothing against big cities, just as long as he doesn't have to live in one permanently. Theoretically he could have moored his boats at the fashionable Aker Brygge in Oslo, but try as I might, I can't even picture this.

We switch to a topic of more current interest. We've both seen the news about the record-sized Greenland sharks that have been caught off Andenes, not far to the north of us, in a southwestern fjord. A Dane caught one weighing 1,940 pounds—using a pole. A Swede managed to bring one up that weighed 1,235 pounds—from a kayak! He told the press that he'd dreamed of catching a Greenland shark ever since he was a child.

"What's so special about that?" asks Hugo.

"Well, when the Dane finally brought the shark to the surface, he started to cry, comparing the experience to a religious revelation. He'd brought along divers with underwater cameras, an assist boat, and a helicopter in order to document the catch. Kind of pathetic, don't you think?"

Hugo merely snorts. Neither of us wants to waste much time on the news of a rich Dane who happens to be obsessed with sharks. In fact, we quickly move on. In one of the deep fjords outside Stavanger another Greenland shark was caught, weighing about twenty-four hundred pounds. Based on the photographs and our knowledge of the anatomy of Greenland sharks, Hugo and I are both skeptical about the report. We don't get the point of such tall tales and boasting, especially when it's so easy for experienced eyes like ours to spot lies.

The video, which was posted on the Internet, shows a Greenland shark that is lethargic, almost comatose. In my mind, the shark was pulled up too fast, and its blood filled with bubbles of nitrogen, like a shark version of the bends. Hugo strongly doubts this idea. He also thinks that the two guys who caught the "Stavanger shark" had no real idea what they were doing. One of them actually jumped into the water to swim alongside the shark.

"If the Greenland shark had suddenly lashed out, as we know it can, that guy would have had the biggest—and final—surprise of his life," says Hugo.

He once saw a film in which Greenland sharks wolf down big chunks of a whale carcass on the seafloor. They keep striking at it, rolling the huge body around, sort of like crocodiles do, until the blubber comes loose. I add a footnote about the cigar shark that lurks in the waters off Cuba. It can suddenly shoot up from below and sink its teeth into the blubber of a dolphin, whale, or shark, and then start rotating. For decades marine biologists wondered what was making the round, symmetrical flesh wounds they'd seen, until someone finally caught the predator on film.

Hugo has found new information on the Internet indicating that a Greenland shark may have attacked a person. In 1856, a human leg was found inside a shark's stomach near Pond Inlet, on the northeast coast of Canada. Of course, the leg could have belonged to a drowned fisherman, or to a passenger or crew member on a capsized boat, or to a suicide or murder victim— practically anything was possible, but nothing could be verified. And there are old Inuit legends that tell of Greenland sharks attacking kayaks.

A legendary encounter between a Greenland shark and a human took place in the Kuummiut region of eastern Greenland in 2003. Dressed in oilskins, the crewmen from an Icelandic trawler, the *Eiríkur Rauði,* stood in shallow water, which was teeming with fish guts and blood. Up on deck the captain spotted a Greenland shark swimming toward the men. The captain was Sigurður Pétursson, nicknamed the Iceman because of his fearlessness. Pétursson threw himself into the water and grabbed the shark, then dragged it far up onto the foreshore, where he killed it with a gutting knife. Afterward the Iceman claimed he had been afraid the shark would set upon his crew. This incident should be classified as a human attacking a Greenland shark, and not the other way round.[1]

At the very least, we know that Greenland sharks aren't exactly picky about what they eat, and they're more than capable of eating humans, if the opportunity arises.

—

Toward evening the air outside becomes clear and cold, and everything seems to expand. The ice appears to be coated with a thin layer of "flour," as my grandfather would have said. It's hoarfrost. The sky is a deep blue. On the horizon in the west the colors change to yellow, red, and purple near the mountaintops. On the highest peaks, light from the sun is faintly visible, like the reflection from a distant fire.

Otherwise the light is blue. Even the snow looks blue.

This intense, and yet muted, winter light is easy to see in Hugo's paintings. He's an abstract painter of the subtle light in the "dark" season. He draws a great deal of inspiration from his surroundings, turning them into something either almost unrecognizable or hyperfamiliar, depending on the eye of the beholder.

Dinner is fried *skrei* (cod) tongue with grated raw carrots and an excellent sour cream sauce made with Hugo's special homemade curry mix. This is the season when the *skrei* spawn in Lofoten. Some of the tongues are as big as fish cakes; they must have belonged to cod weighing sixty-five pounds. Hugo's grandmother in Svolvær had a different recipe. She used to boil the tongues in a white sauce. These didn't sit as well with Hugo, and to this day he hasn't been able to shake off traumatic memories about his grandmother's boiled cod tongues. On the other hand, he could probably eat fried cod tongues several times a week.

As we enjoy our meal, we talk about the Lofoten fishing season, which is going on right now. The waters all around Skrova are teeming with fish. Outside Senja and Vesterålen, where the cod have come to spawn on their journey south from the Barents Sea, the fishermen are reporting record numbers of *skrei*.

Huge shoals of the fish have rounded Lofoten Point, and the fishing off Skrova right now is probably better than anywhere else in the world. And that's no exaggeration.

The *skrei* are practically lining up to spawn. And in turn the fishing boats are lined up to deliver their catch to Ellingsen's Fish Landing Center on the other side of the bay. The boats are heavy with *skrei*. Their gunwales are barely above water when they dock at Skrova. Many thousands of *skrei* have been gutted and hang on the drying racks. From the Ellingsen station great quantities of cod livers float past Aasjord Station. Using a fine-mesh landing net, Hugo has fished up a barrel of livers. He'll use the oil to make paint.

During Lofoten fishing seasons of the past, the bay was packed with fishing boats, bait boats, salting boats, and transport boats. The local population doubled and for a couple of months Skrova actually became a small town. However, beginning in the 1970s, more and more fishing stations closed. The reasons are many and complex. Profits were decreasing, but there had been bad times before, without causing closures. Fisheries have always endured natural fluctuations. The 1970s brought some extremely poor cod fishing seasons. During these years, what was referred to as a "seal invasion" also occurred. The seals did eat huge quantities of fish, but they probably didn't deserve the blame. The main culprits were closer to home.

Herring, halibut, and rosefish had also been overfished. Factory trawlers had depleted the cod stock in the Barents Sea, and all quotas—not just those of the trawlers—had been severely reduced. The miserable catches made 1980 a particularly difficult year. Those who depended on fisheries lost a lot of money. The authorities could do nothing but cut the quotas even more, to avoid a total depletion of the stocks, which is what happened in Newfoundland. After just a few years, a feeling of gloom pervaded many coastal communities in northern Norway.

The Aasjord family's location out in the sea had always been beneficial, but it soon became a disadvantage. Oddly, the future of fishing was not on the sea but on land. To survive, stations needed a direct connection to roads—something that Skrova will never have, since it lies close to six miles out to sea from Austvågøy.

For a long time the Norwegian authorities had wanted to turn the smaller-scale fisherman into a farmer or factory worker. Viewed from the power center of Oslo, way to the south, the coastal fisherman had become an unwanted individual, with an unreliable income and perhaps erratic state of mind. In 1937, Bishop Eivind Berggrav had already become a vocal part of this tradition when he wrote in support of actions that were meant to turn the northern Norwegian fishermen into individuals of "more stable mind." He said it would take generations.[2] And today there's hardly a single fisherman left who lives permanently on Skrova.

It was as if the very zeitgeist were conspiring against places like Skrova, which had played such a central role as long as almost all transport took place by boat along the coast. From time immemorial the shipping lanes had been Norway's highways. The national network of roads changed all that, and old island communities that used to be right on the main marine highway were doomed. New centers with more convenient locations were planned. They were situated farther inland, often in very desolate spots deep inside long fjords, in places where only a few houses had previously existed.

Strategists and planners in Oslo redrew the north, in the name of modernization. In the fisheries it became all about industrialization, both at sea and on land. Coastal fishing, which for a thousand years or more had followed the seasons and the natural fluctuations in resources, was suddenly depicted as a national burden—outdated and lacking the economy of scale that characterized the steel industry and factories in central Norway, where shift work kept operations going around the clock. Those who defended small fishing vessels and the traditional way of doing things were dismissed as "romantics." Yet the modernists promoted equally romantic notions about trawlers and industrialization. New fillet factories were established in a few cities and towns, such as Tromsø, Hammerfest, and Båtsfjord. But these vast and expensive structures—dependent on seagoing trawlers bringing enormous catches to land—soon proved useless when the fish disappeared.

Several weeks before I arrived in Skrova, I was at the legendary Lopphavet in Vest-Finnmark. There the fishermen get ice from the Øksfjord glacier, which calves into Jøkelfjorden. The municipality's coat of arms shows a cormorant on a golden backdrop. Their motto is "A sea of possibilities."

In the past there were fish landing centers on every other promontory in Loppa. The fishing banks are still extremely rich, but today there isn't a single landing center in the entire municipality. The folks of Loppa have lived off fishing for many thousands of years. Now they have largely lost the right to harvest the sea's resources, because the quotas have become unaffordable commodities and objects of speculation by investors. The locals have little or no share in the profit others are making in their region.

The last time Hugo was in Barcelona, he went to the fish markets and saw a variety of cod, including cod tongues preserved in the most fanciful ways. But they were all from Iceland.

We decide to catch some cod for ourselves. The RIB has been brought ashore for the winter. And besides, it's not really the right kind of boat for fishing, since it's an inflated rubber boat. We will be using a lot of hooks, and an ill-placed fishhook could cause the boat to lose its breath. Instead, we're going to take out Hugo's fourteen-footer, a small, open boat made of plastic. But it too has been on shore since the autumn. It—and we—have a rather large problem. It has a leak in its hull. The channels, which should hold only air and create buoyancy, have filled with rain. And in the cold, the rainwater has unfortunately obeyed the laws of physics and frozen solid.

"The boat isn't really on its best behavior," says Hugo as I quietly think to myself that we're going out in the Lofoten Sea in a boat that, to begin with, is much too small. Right now it also has minimal buoyancy, and I can't say I like that idea. The ice will eventually melt, but it's way below freezing outside and only a couple of degrees above freezing in the water, so that will take days. Nevertheless, the sea is swarming with cod, and we agree to make an attempt, provided the weather and the sea remain calm.

"If the boat isn't seaworthy enough, we'll turn around."

I nod, look out the window, and don't say a word.

We leave the boat in the water and head in for the night.

22

The next morning my phone rings and wakes me up. It's an elderly gentleman I met two months ago in an antiquarian bookshop in Tromsø, the major city in northern Norway. He was taking care of the shop for the owner. I happened to mention that I was interested in Greenland sharks and that a friend of mine and I were trying to catch one. The man and I did not exchange phone numbers, but somehow he got hold of mine, and now he's calling to give me some advice. Turns out his brothers once fished for Greenland sharks in the Arctic Sea back in the 1950s. His hottest tip to me: "Put a bunch of rotten herring in a fine-mesh bag, for example a net for oranges, and stick your hook in among the herring." He makes me promise to call him if we catch anything, and he wishes me luck.

Hugo doesn't like the fact that other people know about our plans. He thinks they'll laugh behind our backs if we don't catch a shark, and he's probably right. When strangers from six hundred miles away call to hear how it's going, we're not exactly operating with the greatest secrecy.

Outside, Skrova is dusted with powdery snow. Ice crystals glitter in the pale sunlight, making our optic nerves vibrate. It's rare for snow to be scattered so evenly across the island like this. Normally the snow is swept away by the wind, or it melts in the rain brought by a passing low-pressure system.

It's picture-perfect, straight off a postcard. There's a childlike simplicity to it. When kids draw pictures of the world, they often use bright colors, with simple jagged lines to depict mountains. Maybe they'll add some scribbled green grass or a patch of blue for the ocean and then finish with a couple of old-fashioned houses. And right now that's exactly how it looks in Lofoten. Somewhere in Norway a child is probably drawing this exact scene without even reflecting on it.

We set off from Skrova in the fourteen-footer, leaving from the back side of the island. Water flows into the bay from both directions via a small channel that passes between Risholmen and the largest of the Skrova islands. We've brought along a bottle of water to share, two chocolate energy bars, a hand-line for each of us, and the current issue of *Lofotposten.* The newspaper reports that yesterday a *kaffetorsk,* or "coffee cod," weighing ninety-seven pounds was caught off the nearby village of Reine. A "coffee cod" is the local name for any cod weighing over sixty-five pounds and caught during the spawning season. Since the 1970s, *Lofotposten* has offered two pounds of coffee as a reward to anyone who brought in a cod of this size. Today's newspaper also has a story about the annual cod parade, in which the children of Skrova walk through the streets dressed as *skrei.*

Today the sea is not quiet. We know that even before we reach the seaward side. But it's not outright hostile either. The fourteen-footer sits noticeably low in the water, for obvious reasons. Luckily the sea isn't too rough, just long breakers that won't put our heavy, listless boat to the test. Of course that could all change, and much faster than the time it would take to get back to harbor.

The boat could serve as a deep freezer. It probably contains enough ice for two thousand cocktails, and four thousand whisky on the rocks. A couple of drinks would be great right now, to stop me from worrying about heading into the open Lofoten Sea in this frozen boat.

The seagulls are silent, the snow sparkling white. Even the sun seems cold. For me, coming from the big city only yesterday, the dazzling clear surroundings and the wide horizon are refreshment for my soul. Yet there's something about the sea today that has me concerned. What could be lurking behind the silvery-white and viscous fluidity? It's like staring into a glass eye.

Hugo spots a bunch of small commercial fishing smacks a good distance away in Vestfjorden, and he sets course in that direction. They have echo sounders, and underneath the boats the *skrei* are bound to be plentiful. I like the plan, especially because at least then someone will be around to pull us out of the water if necessary.

After fifteen minutes of chugging along, our thirty-horsepower outboard motor gets us out to the fishing grounds. By the way, did I mention that the outboard is actually bigger and heavier than the boat is certified to carry, making the center of gravity wrong from the start?

We take up position an appropriate distance from the fishing smacks but close enough so we can see the nets and lines bringing in big, strapping *skrei*. All we have to do is lower our hand-lines, which have strips of brightly colored rubber attached, doing a not very impressive job of hiding the hooks. But it works. The *skrei* are about 130 feet down, and as soon as the hooks reach that depth, the cod start biting and we pull them up.

The *skrei* are at this particular spot due to the temperature. They like to be where the warm layer of the deep and the colder water closer to the surface meet. The Norwegian scientist Georg Ossian Sars was the first to discover this.

In 1864, Sars traveled to Lofoten to study the biology of the *skrei*. With Skrova as his home base, he was rowed around in Vestfjorden, presumably by people who lived on the island. In Skrova a little park is hidden away and might easily be mistaken for the lawn of a nearby house. But at this spot you'll find a monument, put up in 1966 by scientists at the Institute of Marine Research and the Norwegian Department of Fisheries, to honor G. O. Sars. Etched into the stone, it says that from Skrova he had "elucidated the most important traits of the cod's biology."

During certain times in the spawning season, the *skrei* eat very little. Fishermen say that the fish are "moping" and use nets instead of hooks. But not much moping is going on below us today. The biggest *skrei* that we pull up weigh thirty to forty-five pounds. One of them close to sixty-five. Some have swallowed the hooks, but many have the hook affixed to the outside of the mouth, in the eye, or in the side of their body. They have to be hauled up horizontally, and it takes quite an effort to get them to the surface.

It's hard to ignore the fact that the distance between our boat's gunwale and the sea is less than desirable, and diminishing. Even the crews on the fishing smacks look startled when they suddenly catch sight of us at the top of a wave, before we disappear into the trough. Some of them shout and wave, and we wave back. Maybe they think we're in need of rescuing.

We're not. At least not by our own definition. We focus on hauling in the *skrei*, which are swarming in big shoals down below.

Under such circumstances, it takes a special kind of person
to decide to pack up and head for land, even though the boat
is small, without buoyancy, and about to be filled with several
hundred pounds of extra weight. We're not those kind of people.

For a few days, female and male *skrei* swim close together, with
the males moving sideways. Roe and milt are sent out simultaneously from the male and female. Then the *skrei* use their tails
to whirl the two together in order to fertilize the eggs.

The *skrei* we pull up are bursting with roe and milt; in other
words, they haven't yet spawned. They will very soon, and when
they do a couple hundred *trillion* (not billion) cod eggs will be
floating around in the Lofoten basin. A female cod can have up
to ten million eggs. Not all will be fertilized, of course, and a lot
can go wrong. In the beginning, the cod larvae live off their own
yolk sac. The eggs that drift around may get destroyed or eaten
by others. After a couple of weeks, when the eggs hatch, most of
the spawn meet the same fate. They try to grab plankton—first
phytoplankton, then zooplankton and krill. When the small,
transparent fish are four weeks old, they leave the upper layers of the water. From then on, they will try to survive on the
bottom as they drift north with the Gulf Stream toward the
Barents Sea.

The first year is the most dangerous. After that, cod face few
threats.[3] Cod that survive for seven years are ready for the long
journey back to Lofoten, where they will spawn. Of the many
millions of eggs from each female, at least two individual cod
must survive in order for the stock to remain stable. No one

can say exactly why this year in particular seems to be a banner season, with many hundreds of millions of *skrei* swimming below us.

Most people know that the world's richest cod stock spawns off Lofoten and Vesterålen. But those same areas of the ocean are also very important for halibut, which also spawn in the winter, and herring, which spawn in the spring. Then there are the stocks of rosefish, pollock, haddock, wolffish, and angler. Traditionally, Lofoten has millions of seabirds as well, just like other places near the sea in Norway. But in many instances, the numbers have decreased to an alarming level. Many types of fish, on which the birds depend for food—such as the sandlance, capelin, blue whiting, and Norway pout—have been subjected to overfishing. Not as food for humans, but for farmed salmon.

You might be surprised to know that cod and Big Oil like the same thing: plankton. While the cod eat the plankton fresh in the sea, the oil companies prefer it to be two hundred million years old and transformed into sticky black fuel. The new Norway depends on this oil, just as in the past we depended on cod, cod oil, and herring oil. In the old days, the fishermen would toss oil on the water to break up the waves when they were trying to rescue the crew from a sinking ship. Today commercial trawlers throw fish into the sea. One thing is for sure: the world's richest spawning banks are threatened by oil. If it comes to a blowout, the Lofoten Wall is at risk of acting like a long natural oil boom, trapping the oil along its beaches and killing off everything from seabirds to plankton. Even tiny amounts of oil can destroy fish spawn.

If Tanzania started to drill for oil in the Serengeti, the whole world—and presumably with Norway at the forefront—would protest. We would find it barbaric, and maybe we'd even donate a billion kroner to stop them. Norway already doles out billions to save the rain forests in Brazil, Ecuador, Indonesia, the Congo, and other places in the tropics. Yet Norway has an equally unique area, an underwater Serengeti. In this place of unparalleled fecundity, which is world renowned for its beauty, Norway, in spite of being one of the richest countries, wants to start drilling for oil.

Melville's subterranean miners keep on working.

As we drift, hauling up *skrei,* I tell Hugo that in the 1960s, Soviet marine biologists developed a theory that the sperm whale uses its enormous sound organ as a weapon, an "ultrasonic projector" or "sound laser." The idea was that highly concentrated and precisely aimed sound waves enable the whale to paralyze squid and other prey. American scientists have followed up on this research, hoping to put it to military use.

Like the Greenland shark, the sperm whale catches animals that are much faster (squid can reach a speed of thirty miles an hour), and it often does so in total darkness, way down in the deep. But until recently no one had ever observed the sperm whale in action. Close to the beginning of the new millennium, Danish whale researchers investigated the theory off the Norwegian island of Anøva in the Vesterålen archipelago. Using advanced hydrophones, they discovered that the clicking noises made by sperm whales were focused and could, to a great extent, be directed at specific targets.[4]

Before the world's oceans became filled with noise from propellers and machines, whales could hear one another at a distance of about six hundred miles.

Over the past few years, because of possible oil, a lot of seismic surveys have taken place near Andenes, Vesterålen, and Lofoten, as well as in many other places farther north. Sometimes called "seismic shooting," this involves sending shock waves through the ocean. Active fishermen on Andøya think this is why Vestfjorden—in fact, the whole region—is teeming with mackerel. Seismic shooting keeps away minke whales, pilot whales, killer whales, and other mackerel eaters.

Coastal fishermen, environmentalists, and whale researchers fear that the shock waves may injure or kill whales—and maybe also fish spawn. They point out that whales do not behave normally in areas where shooting goes on, because it may damage their ears. For whales, the sound waves must feel like sonic carpet bombing. After all, the sound has to penetrate several layers of mountains on the seafloor.[5]

Hugo shakes his head, as if to say that he's lived long enough for this not to surprise him. He saw on the news that the bodies of twenty-six dead pilot whales had washed ashore in Vikna municipality in northern Trøndelag while a seismic survey was going on in the ocean offshore.

I happen to think of something else I read recently. In the 1950s, an American scientist, Dr. Harry Wexler, came up with the idea that the earth—or at least the United States—would benefit if the North Pole became free of ice. Global transportation would be easier, and arctic raw materials would be more accessible. Wexler proposed detonating hydrogen bombs under the polar ice cap. Maybe only ten would be needed, each around ten megatons. This would create enough steam to encapsulate the entire North Pole in a thick mantle. And then the ice could no longer reflect sunlight. The heat would become trapped—the greenhouse effect was already well known—and the rest of the ice would melt.

Hugo looks at me as if he thinks I'm joking.

23

By now we've slipped into a rhythm and we haul yet another big, wriggling cod up through the water column. Then we hit it on the head with the gaff and lift it on board the fourteen-footer, where we swiftly stick a knife in what the old folks used to call *kverken,* the "craw."

Each cod hauled onto the boats crowding the Lofoten Sea has been swimming for years over several thousand miles. Right now all that matters is spawning, but the ones we catch are stopped just before they reach the finish line. The fish that takes our bait may not understand what is happening, but it does have a nervous system. What a shock it must be to find itself suddenly nabbed and then dragged up toward the light by an invisible force. Yanked way from the other fish in the shoal (do they notice it disappear?), pulled from a depth of 150 or 200 feet all the way up to the surface. Of course the fish struggles with all its might, and it might even manage to wriggle free (does it feel relieved?). But most get knocked on the head by a gaff and are then hauled over the gunwale into a boat where many other fish have already suffered the same fate. Do they have any biological or intuitive understanding that they're about to die? Or do only more evolved animals have that sort of awareness?

One more fish, and then another. For us, it feels equally great each time, and that's the problem. Serious numbers of *skrei* are swimming below us. And serious numbers of *skrei* are beginning to fill up the boat.

Hugo tells me that in the old days cattle were fed protein-rich *skrei* milt and roe, if it could no longer be used for caviar. He also says that the Japanese and some Lofoten inhabitants drink the milt as a cocktail or aperitif, called a *krøll*. Hugo feels nauseated by his own story, but of course he's unable to vomit.

The sea renews itself with each moment. Everything glitters. As usual, everything is in motion out at sea. When we started fishing, the rhythm of the waves was steady and calm, like the breathing of a huge, slumbering creature. Now the swells are intermittently breaking up, getting choppier and rougher. Our boat turns its stern seaward, and the tongue of a wave reaches over the gunwale and into the boat. Something is about to change. Black patches are hovering over the open waters. It's a strange sight, because in other places the sun breaks through the cloud cover in vertical shafts of light that sporadically disappear as the weather system moves around, like in a cartoon or on the stage set of an opera.

I don't know whether Hugo has noticed, but for the first time in all the years we've gone out to sea together, I don't feel safe. The RIB we usually use cannot sink. Not completely. Even if all the pontoons get punctured, the hull will more or less stay afloat. The fourteen-footer is a different matter.

In general, Hugo feels perfectly at home out at sea, and he knows these waters like the back of his hand. He has also survived the most unlikely situations on the ocean. And since it has "always ended well" for him, maybe that's why I'm thinking right now he might have become a little reckless. You need only one exception to the "always ended well" pattern. It suddenly occurs to me that the stories we tell are those that end well, meaning the person is still around to tell the tale. Could this be the one time when things will go badly? Will this be the story others will reluctantly have to tell?

I once asked Hugo when he came closest to drowning. He told me a story I'd never heard before. When he was about twelve years old, he and a friend rowed out to a small islet in Steigen, just to explore it. While they were on the far side of the islet, the tide took away their badly moored boat. Hugo swam after it, but the current was strong and went in the wrong direction. The little empty boat was transported out to sea faster than Hugo could swim. He was closing in but not fast enough, and suddenly he was too far from shore to be able to return. His powers were drawing to a close, and he stopped the chase, totally exhausted. He was preparing to meet his Maker when something gently touched his foot. Hugo caught it. It turned out to be a rope. Not just any rope but the long rope hanging from the boat. With the last of his energy, he was able to reach the boat and pull himself aboard.

In Norse mythology, the goddess of the deep sea was called Rån or Ran, meaning to rob. With her net she would catch—no, steal—drowned seamen. Rån was married to Ægir, brother of the wind and fire who wore a crown of seaweed and ruled over the oceans. Their nine daughters represented the nine waves of the sea and were named accordingly. According to the Old Norse poets, a boat that sank would disappear into Ægir's jaws, while Rån took the crew to her deep-sea realm. Ægir ruled over both calm and stormy seas. He brewed the mead of life from Balder's blood, and his goblet refilled of its own accord. For the Vikings, Ægir symbolized prosperity, but not merely because he had access to unlimited amounts of mead, or because he and Rån lived in a golden palace. These luxuries were just a tangible manifestation of something else: the unfathomable wealth of the sea.

Our boat is what folks in the old days might have called a floating coffin. At least we're wearing flotation suits. And that's a good thing, but only to a certain degree, as Hugo explains when I nonchalantly ask him about our clothing. He emphasizes that what we have on cannot be considered a *survival suit*, and then he stuffs a piece of cooking chocolate into his mouth. Whenever he goes out in a boat, he always takes cooking chocolate and hazelnuts along. They are essential items, serving as emergency food rations. One side effect of his botched stomach operation is that he sometimes runs out of energy. All his strength vanishes, and he can literally no longer stay on his feet. This has happened only a few times but always at the most inauspicious moments. The last time, he was out hunting rabbit in the wooded area behind his house on Engeløya. He came crawling home, making his way on all fours over the grass and up to the porch, dragging his rifle behind him. Sweat poured off him, and he couldn't utter a word, but Mette realized that he needed to eat something. In the kitchen there was a platter of herring, and in ten minutes Hugo downed eight pieces of bread topped with smoked herring. A normal meal for him would be one sandwich.

Sometimes even a survival suit isn't enough. A couple of

years ago, the body of a man floated into Svolvær harbor. He turned out to be a fisherman from Melbu who had gone missing a while back. Wearing a survival suit, you can stay alive for a long time, depending on the season of the year and what you have on underneath the suit. I wonder what the man was thinking when his fishing boat was about to sink and he put on the suit. He probably assumed everything would be fine. But that's not how it turned out. Presumably his fingers got too cold, so he was unable to pull up the zipper the last inch. Water quickly seeped inside the suit, and he was doomed.

It's a fine line between life and death. At about the same time as this incident, a sixty-six-year-old fisherman was out in his boat when he developed engine trouble. The anchor failed to grab hold, and the current began carrying the boat straight toward the rocks. The fisherman was only wearing ordinary clothes and a life jacket. But before he went into the water, he somehow managed to put in an SOS call on his cell phone and gave his approximate location to the emergency operator. His boat was now very close to the rocks and about to be smashed to pieces. A strong wind was blowing. It was fourteen degrees Fahrenheit and pitch-dark when the fisherman had to throw himself into the ice-cold water. After the waves had pulled him under several times, he finally managed to scramble up onto a small, slippery rock and hold on tight. Twenty minutes later a Sea King rescue helicopter from the 330 Squadron in Bodø arrived. Using spotlights, the crew was able to locate the fisherman. They lowered a rescuer down with a basket to get him. By then the fisherman had lost all feeling in his fingers, and the rest of his body was also going numb.

Only a week before I arrived on Skrova, an elderly man was

found drowned near the island's eastern inlet, with his empty pleasure boat circling nearby. He'd been out fishing, and for some unknown reason he fell out of the boat.

Fishing is without a doubt Norway's most dangerous profession. No one knows exactly how many fishermen have drowned during the Lofoten fishing seasons, which started long before Harald Fairhair united Norway during his reign as king from circa 872 until 930. But to give you an example, in 1849, more than five hundred fishermen were reported drowned in a single day when a violent storm suddenly took hold. Over the course of any season several thousand people could lose a father or husband, and in those times, a primary breadwinner.

If we combine the official records of the Lofoten Supervisory Force for the years 1887–96, we find that 240 fishermen drowned as a result of "shipwreck." According to reports, the main reason that boats sank was because they were either deluged or capsized by a big wave.[6] It's the primordial equation—almost a mathematical given. Overloaded boat + big wave + cold water = drowning.

I begin to muse aloud. "Just think how many fishermen have drowned during the Lofoten fishing season over the years. Five thousand? Twenty thousand?"

Hugo considers this for a few seconds before replying.

"And who knows—maybe a Greenland shark came along and gulped some of them down as they struggled in the water."

I scan the sea. As long as there are other boats nearby, we're safe, I remind myself.

"So what do you say? Looks like we might have enough, don't you think?" Hugo asks me.

Half the boat is filled with *skrei*. We have to wade through the fish every time we move.

"Are you sure?" I say sarcastically as I bail water and *skrei* blood out of the boat, using an old paint can that Hugo has designated for the purpose.

"Let's pull up the bait line," says Hugo.

I check my cell phone. It's running out of juice, but the battery will probably last another hour. I've taken off my mittens to bleed the fish, and my fingers are slimy and ice-cold. Even though it's nowhere near five degrees Fahrenheit, like it has been the past few days, it's still freezing. My phone slips out of my grasp like a bar of soap and lands in the bloody bilgewater. It could just as well have fallen into water twenty-five feet deep. Hugo checks his own phone. It still has a little charge left.

The waves are now higher than when we set out. There's no doubt about it. The translucent, almost crystal-clear day begins to change its appearance. Hugo peers out at the open ocean, his gaze lingering a bit on the horizon. It looks as if a curtain has been pushed aside, and thick cigar smoke is trickling in our direction. Hugo starts up the outboard motor and sets course for Skrova.

"It's going to snow," he says. The motor coughs as it attempts to pick up speed. The boat is so weighted down that it seems like we're hardly moving.

A few minutes later heavy, wet snowflakes begin to fall. We're far out in the fjord, in the middle of what's called a *rennedrev*, a combination of stormy weather and snow.

Our safe and familiar points of orientation—Skrova and the surrounding islands—instantly blur. The Skrova lighthouse isn't much help now. The world has turned monochromatic. A snow squall darkens the sky. Everything seems to close up around the boat like a sack being laced up.

"This is not exactly great," says Hugo in that odd understated way of his, putting emphasis on the last word, as he continues steering the boat more or less blindly. He knows that even if we veer off course, there's still a long way before we approach the deadly skerries and shoals. On the back side of Skrova, which is where we're headed, there are places among the reefs that are particularly treacherous—in Norwegian, such places are called *støvelhav,* or "boot sea," because of how shallow they are.

For a moment visibility drops to zero. Then we catch a glimpse of an island. But which one is it? It feels as if they've started moving, shifting shape and position between every crack we see. Over there I thought I caught sight of Lillemolla, with its sharp peaks, or was it the top of Skrova? Now, in that same direction, I see what looks like a small islet, but I don't recognize it at all. The world is in flux, distorting perspectives, as if viewed through the glass of an old windowpane. If Schoenberg's works of music were transformed into images, they might resemble this counterpoint scene.

Our boat is weighted down like an ice-coated tree branch right before it snaps. All of us will eventually die, but those who disappear at sea are truly gone, suddenly and forever. *As if sunk into the sea,* becoming part of it. Long ago I had a friend, or rather an acquaintance, who got his foot tangled in the line while the trawl net was on its way down into the deep. His body was never found. That was thirty years ago, but I still think about him. My great-great-grandfather drowned at sea, but that's not a family tradition I'm eager to continue.

—

The deep, salty black sea rolls toward us, cold and indifferent, lacking all empathy. Detached, merely itself. This is what the ocean does every day. It doesn't need us for anything, it doesn't care about our hopes and fears—nor does it give a damn about our descriptions. The dark weight of the sea is a superior power. Many have been in this situation, ever since some of our over-confident ancestors set a hollowed-out tree trunk in the water and paddled off on languid waves, only to venture out too far, where the currents were stronger than their arms or paddles. Or maybe, like us, they were surprised by a storm. All of them must have felt the same cold shiver when they realized the sea is truly without sentimentality or memory. Whatever it swallows is gone, becoming food for the fish, crabs, and annelid worms, for the lamprey, hagfish, flatworms, ringed worms, and all the parasites of the deep. To be drowned and embraced by the eternal, indeterminate All.

When God wanted to punish Jonah, he sent a huge "fish" to swallow him. Jonah cried for mercy as the depths of the sea surrounded him from all sides. Inside the belly of the whale the water reached up to his neck and seaweed encircled his head. But the Lord merely wanted to teach Jonah a powerful lesson, so He made the whale bring him up from the realm of the dead and spit him out on land. Fear turned Jonah into a faithful believer in God. Even Islam respects the whale for this reason. In the Koran, it says the whale that swallowed Jonah is one of the ten animals that will enter heaven.[7]

It's a real williwaw! I remind myself that in the old days, fishermen were always getting caught in this sort of situation. And in boats that may not have been any bigger or more seaworthy than ours. They used sails, and yet these gritty and skillful men always mastered the conditions. But wait a minute—that wasn't what they did at all! No, they drowned by the hundreds, nearly every single fishing season, at precisely this time of year and in this same area. In Skrova there is an old local song about the sea opening its treasure chambers so generously:

> *But suddenly it turns in furious anger*
> *demanding what it had given, plus interest.*
> *Oh yes, all that is left may be only*
> *scraps of what was once a boat.*
> *The sea can give, but it can also take;*
> *the crew remained in their wet grave of kelp.*[8]

I give Hugo a surreptitious glance. He doesn't look worried. On the other hand, have I ever seen him look worried when he's out at sea? At least he's not still wearing his headset. What happens now if the currents displace a wave so it merges with another one to form a swell double in size—a breaker, in fisherman's terms?

The bottom of the boat is completely covered with *skrei,* their gills—or *toknan,* as Hugo calls them—still moving. When the fish swim around, they make certain sounds. Cod, for instance, utter grunts or issue a series of deep tones, reminiscent of Morse code patterns. We can't comprehend their language, but these dying *skrei* seem to be telling me something.

Beneath us in the dark, the water moves restlessly over the sandy bottom and smooth rocks. Even the starfish on the seafloor have to hold on tight. The fingerlike fronds of the sea girdle shift back and forth, swaying like tall grass in a strong wind. The halibut is in control, moving calmly down to deeper waters. On the bottom it slips into the sand, like pulling on a dressing gown, and settles in. Spawn of cod, pollock, haddock, herring, and mackerel try to steady themselves in the restless kelp. The Greenland shark, semi-blind, lies in the dark, so deep in the water that it hardly notices what's happening on the surface.

Hugo reduces speed and asks me to keep a lookout. As long as we can't see anything and the boat continues to be carried forward by the current, we have a problem. To move in these waters we need to keep track of the many shallows and skerries, but we have basically no idea about our position. Hugo, of course, is fully aware of this. For him the sea is something different than it is for me. He's much better at reading the fleeting glimpses of visibility. And even though he can't see land, the ocean is not a uniform, undifferentiated element with no identifiable features. Every position at sea is like a location in a landscape—a *place*—with unique currents, specific conditions on the bottom, various shallow areas, and other significant features. But you need skill, and at least some visibility, to see them.

Neither of us says much, but every once in a while Hugo asks me what I think. Was that Lillemolla we were catching a glimpse of right now? Land and sea still seem to be changing places. He merely asks me for the sake of appearances, because in this situation he trusts his own judgment best. And I put my trust in him too, since I'm completely disoriented. All I can do to help is shout if I see anything right in front of us. When the snowfall is at its heaviest, the flakes make it hard to keep my eyes open. Through narrow slits I can barely see even a couple of boat lengths ahead. The snow is a threatening wall that erases all contours. My biggest worry is not that we'll strike land, but that we won't. Because now the wind has picked up, and the waves have too. It never ceases to amaze me how quickly the wind seizes hold of the sea.

The fourteen-footer seems smaller than ever, and the ocean seems much bigger. The boat, Hugo, and I are stone-cold sober. It's the sea that's drunk. How many times have I leaned over the rail to stare down into the abyss? Now it's staring back at me. In the Skrova song, a couple of lines are devoted to this feeling: "Storms and rough seas are crushing forces / man is but a seed."

For once Hugo hasn't brought along any line or anchor. They're in the RIB. I ask Hugo if we have enough gas in the tank. He frowns, checks, then nods. He has gone unusually quiet. He's hyperalert and locked in, almost like he received an anonymous threat and is trying to assess whether to take it seriously or not.

Sitting in the bow, I get drenched with sea spray, so I decide to move toward the center thwart. My movement shifts the balance in the boat. Hugo always sits in the back with his hand on the throttle of the outsized outboard motor. Just as I'm starting my move, a big wave strikes. The fish crates slide to the back of the boat as the water washes over the stern. Hugo sets his feet against a fish crate and kicks it away with all his might, then hurls himself after it. With so much weight in the wrong place, the boat could have filled with water and gone down in an instant.

Sheepishly, I creep back to my post at the bow, with no intention of leaving it again.

—

It's still early in the year, and soon it will be completely dark. With this cloud cover, which stretches across the sky, it's already quite dim. The wind and darkness descend upon us like two allies on a mission, teaming up with the blue-black sea, which is churning around the islets, and the underwater shoals that await us. The thick, wet snowflakes are starting to freeze. It must be getting colder where they're coming from.

Row, row, row your boat, gently down the stream . . . The fourteen-footer is leaping up and down like a carousel horse. There is a clarity in the sinking water, in the vertical movement taking us down into the ocean. It's in front of us, above us, inside us. But most of all: below us. Down at the bottom of the dark sea, where the wondrous fish live.

Suddenly the curtain opens, as if yanked aside. How does the song continue? "Then, through the cloud cover, a ray of light finds / the way to Skrova with joy and hope." We get the visibility we need. Snow-covered islands with jagged black peaks of granite appear as if in a vision several miles ahead of us, on the port side. Hugo knows instantly where we are. We've drifted farther west than we would have thought possible, especially since the wind and sea currents are coming from that direction. If we'd chugged along for another hour, we would have undoubtedly ended up in unfamiliar waters around Henningsvær, or farther west, a long way from Skrova.

Now everything returns to normal. We keep crawling along as we eat some chocolate and take a couple of gulps of water, without saying anything, because certain situations require no words. Twenty minutes later we arrive in Skrova harbor, in the opposite direction from which we left. The fourteen-footer is still full of *skrei*. We didn't have to toss any of the fish out in order to stay afloat. Back on shore, neither of us talks about the trip as any sort of dramatic incident. And maybe it wasn't. Safely ashore, it becomes a trip I wouldn't have wanted to miss.

24

If you've been out fishing in the Lofoten Sea, it's not just a matter of docking the boat and then calling it a day. Half the job still remains. Now you have to take care of the catch. We set up a cutting table on the wharf, and soon fish guts are flying through the air. Hugo cuts out the tongues with one swift Japanese maneuver.

To make proper dried cod—called *rotskjæringer* in Norwegian—you have to remove the backbone so the fish can hang by its tail over a pole, with the boneless fillets draped on either side. This butterflying method is time-consuming, but it produces the best results. Some people hang up whole, cleaned fish, but then there's a risk the belly will close up, which affects the drying process. Olaus Magnus reports that even back in his day the *rotskjæringer* were the most valued dried cod and were used for the most flavorful dishes.[9]

While Hugo does the cutting, it's my job to tie a line around the base of the tail so the fish won't fall because of its weight. We also have to treat the liver, roe, and tongues. We put the roe in a basin and sprinkle it with layers of salt. The roe shouldn't be *gotten,* as it's called in Norwegian. At the stage when the eggs are almost ready to spawn, the roe is too jellylike and fatty.

Fortunately we have only a few of those. As the roe dries, the salt presses out the liquid, and then Hugo will smoke the roe and make caviar. We also salt some of the *skrei,* to dry them as *klippfisk* later on.

We put the livers in a big plastic bucket. Over the course of the next few weeks and months, the livers will separate, and pure cod liver oil will float to the top. Then we'll mix it with paint to use on the walls of Aasjord Station, wherever needed. Left in the bottom of the bucket will be the *graks,* an oily waste substance that smells particularly awful when it rots. We're going to use the *graks* as chum to catch the Greenland shark. Hugo tells me that in the old days they used to compress the *graks* into palettes that were then wrapped around pipes to keep them from freezing in the winter. The material created a heat-producing gas.

The oil from cod livers is superbly suited for making paint. But the paint from the oil of a Greenland shark is in a class all by itself. In Lofoten there are still houses that were painted with the stuff fifty years ago. The paint gets so hard that it's impossible to scrape off, and it's so smooth that no other paint will stick to it. If you want to change the color of the building, you have to change the boards. They ought to paint spaceships with Greenland shark oil, even though the stench would spread through space and give our planet a bad reputation.

While we're busy with all this, I happen to think about what I read in today's *Lofotposten.* It's March 25, also known as "Great Liquor Day." The origin of the name is unclear. This would be around the date when first-time crew members had earned enough to buy their fellow seamen a round of drinks. That's one possible explanation. Another theory is that the tradition goes all the way back to when Norway was Catholic and has something to do with the Feast of the Annunciation, marking the day the Archangel Gabriel announced to the Virgin Mary that she would be with child. It's not known how alcohol came into the picture, but, as they say, God works in mysterious ways. Regardless, I seem to recall that I have a bottle of whisky up in my room. I bought it on the Orkney Islands because they said it was the best "salt" whisky made in Scotland.

Mette comes home with icicles in her hair after taking a swim in the sea. She too has spent all her life in a fishing culture and nods approvingly when she sees us. There are guts, roe, livers, and tongues everywhere, in basins and buckets. With loud slaps, we hang up the gleaming, ice-cold *skrei* on the drying racks as the day's light continues to recede around us. When the time comes, some of the dried fish will become lutefisk—dried cod treated with lye—but not like the inferior kind some shops sell. Lutefisk made from third-class dried fish will dissolve in water, but ours will stay firm.

The drying process has always included an element of chance, because the quality of the dried fish depends on the weather. It can't hang too long in a severe freeze, because then it will split—becoming what the Norwegians call *fosfisk*. Too much direct sunlight isn't good either, because the fish may get scorched. Luckily, the Lofoten fishing season occurs during the two months of the year when drying conditions are optimal. If the *skrei* came to Lofoten later, the temperature would be too warm, and the fish would be ruined by insects, mold, and bacteria. If it was earlier in the winter, the freezing temperatures would stop the drying process, and the fish might turn sour from frost erosion. The fact that for all these years they've been able to produce dried fish in Lofoten is due to a combination of fortunate circumstances. Not only do the fish happen to come to this specific location, and in huge numbers in a good year, but the time of year is ideal for drying.

As for the *skrei* we're now hanging up to dry, we hope for a mild and slightly damp breeze, as well as a lot of light, but no heat—a couple of degrees above freezing—so the fish will dry and mature at the proper pace. A little rain won't hurt, but a lot of rain over an extended period wouldn't be good. The professionals prefer to hang the fish with the backs to the southwest, so that no rain will get inside the belly. The air shouldn't be too dry, either. Warm, stagnant air results in poor quality. Luckily, Skrova is seldom bothered by that type of weather.

If the drying goes well, we'll end up with the most long-lasting, flexible, good-tasting, and protein-rich foodstuff I can think of. Cod is a lean fish, and in dried condition it retains all its nutrients in concentrated form. Through the years cod has been Norway's most valuable export item. In *Egil's Saga,* Torolv Kveldulvsson is said to have exported dried fish from Lofoten to England as far back as AD 875. The oldest verifiable historical sources reveal that the then market town of Vágar on Austvågøy just north of Skrova was the first to export dried fish.

When dried fish sorters evaluate fish for the export market, they consider a number of factors, one by one: color, smell, length, thickness, consistency, and appearance all play a role. Does the fish have gaff marks? Bloodlines and blood spots? Or maybe traces of its own liver on the neck or belly because it hasn't been properly cleaned? Have birds been at the fish? And of course there can't be any trace of mold or mildew. In the centuries that have passed since dried fish sorting was made mandatory by royal decree in 1444, the sorters have developed their own language. Sources from the mid-1700s in the Hanseatic town of Bergen, which was largely built on the production and sale of dried fish, reveal the widespread consumption of cod. Various qualities were described as *lübsk zartfisk, hollender zartfisk, hamburger høkerfisk, lübsk losfisk,* and so on.

The dried fish sorters operate today with thirty different levels of quality, some of which are carryovers from the Hanseatic period. The three main categories are *prima, sekunda,* and *Africa.* The Italians pay the most for the sorting called *ragno,* a thin and flawless fish longer than two feet. The belly has to be open for inspection. All sortings within the categories of *prima* and *sekunda* are initially intended for the Italian market. Other, cheaper sortings often go to Africa.

On the plane to Bodø, I happened to sit next to a Nigerian gentleman living in Manchester. He was a fish broker, and he was on his way to Lofoten to contract for futures, as he said, with the dried fish producers. Dried *skrei* heads, in particular, are highly valued in some West African countries, where they make delicious stews and curries from them. Late in the spring he would sell to Africa the dried *skrei* heads, which at the moment hadn't even been pulled out of the sea.

For supper we have little burgers of *skrei* cheek fillets. You fry the cheeks with the skin side down. This meat is somewhat different from the rest of the body. It's firmer with slightly coarser fiber, and has a shellfish-like taste.

While we eat, Hugo tells me a strange—no, a grotesque—story. When he was a boy, in the mid-1960s, three big pyramidal *hjeller,* or drying racks, were built in Helnessund. Tens of thousands of pollock were hung on the racks in the middle of summer. In northern Norway, pollock are not normally hung up to dry, but these fish were intended for a different market. Civil wars were raging in several places in Africa, bringing catastrophic famine.

But flies got into the fish. So before the fish were exported, men in white hazmat suits sprayed the fish with DDT, a strong, toxic insecticide. Fortunately, as Hugo remembers, the export of dried pollock to war-ravaged developing countries in Africa stopped after a couple of years.

My last thought before I fall asleep with my clothes on is that someone ought to keep watch so mink don't come and help themselves to the *skrei.*

25

The next morning I get a cup of coffee and go out to the wharf. The *skrei* are intact, but an otter out in the bay comes swimming past Aasjord Station, right up to the floating dock. The otter isn't exactly trying to be discreet, because it leaps through the water like it thinks it's a dolphin. Suddenly it stops, rubs its tiny paws together, and stares up at me. Hugo comes out to the wharf, and I point at the otter. After a few seconds it swims away, moving in the same dolphinlike manner. Hugo and I stand there, laughing. He's never seen an otter swim that way, and it seems a little strange for the animal to move like that in the open bay of Skrova, and in broad daylight. Hugo often sees otters when he goes out fishing around Skrova, and they always seem to be having a lot of fun, especially in the winter. They slide down into the sea from steep, ice-covered rocks. Then they climb back up and slide down again. Their behavior serves no obvious purpose; they obviously do it for one reason only: to play. Otters are known to be smart. Sometimes they float on their backs in the water, holding a stone in their paws. They use the stone to smash shellfish against their chests.

The otter is native to Skrova. The mink, however, was brought here from America close to a century ago, to be raised for its fur. Of course a lot of the mink managed to escape, and they more or less adapted to living in the wild. Mink get into everything and don't have much restraint, wreaking havoc whenever an opportunity presents itself. They also kill huge numbers of seabirds.

In the afternoon we go out in our boat, but not very far. It's good weather, so in a couple of hours we get a day's catch of cod. The fourteen-footer is of no use when it comes to hunting for Greenland sharks on the seaward side. That's not something even worth discussing. Of course, it's unfortunate because in my small traveling library, I have a book that gives me good reason to believe that right now the deep is teeming with Greenland sharks.

Johan Hjort (1869–1948) was one of Norway's truly great ocean scientists. In 1900, he set out on a yearlong trip along the coasts of northern Norway on board the "fishery researchers'" new steamship, named for the eminent Michael Sars. Hjort was not only a scientist, but at that time he was also the director of Norway's fisheries. In the north he wanted to record independent observations regarding all the fisheries. In 1902, he published *Fiskeri og Hvalfangst i det nordlige Norge* (Fishery and Whaling in Northern Norway), which I've brought with me to Skrova.

In the introduction, Hjort writes that he wants to shed light on "the big questions that have preoccupied the people of northern Norway, and which have become generally known because of the old conflict between fishing and whaling." Back then coastal fishermen in Finnmark believed that whales normally chased great numbers of capelin toward the coast, but when whaling ships went out to hunt, the whole natural balance was disrupted. Simply put, capelin no longer came to the coastal areas, and for this the fishermen blamed the whalers. Over the course of one season, they might kill up to a hundred blue whales and several dozen fin whales in Varangerfjorden alone. The coastal fishermen also thought that the waste materials from the whaling furnaces and factories polluted the seafloor.

Since Hjort had set out to investigate both the economic and marine-biology factors for all the fisheries, he couldn't ignore the Greenland shark. He admitted that scientific knowledge about the species was far from complete, but he concluded that huge numbers of Greenland sharks existed in the Arctic Ocean. At that time there was a significant focus on fishing for Greenland sharks in the north. In the winter, according to Hjort, the sharks might even venture all the way south into Bunnefjorden, near Christiania (present-day Oslo).

In late winter, large numbers of Greenland sharks were found in the coastal banks off Nordland just as the *skrei* were arriving to spawn. Hjort writes that in order for the *skrei* fishing to get going, the Greenland sharks first had to be chased away, but he doesn't say how this seemingly impossible task was executed. In Finnmark alone—especially from Hammerfest to Vardø—the shark was hunted from six ships and twenty-one (motorized) vessels during the years Hjort visited. His description of how the shark was caught shows that Hugo and I aren't entirely off the mark. I go to find Hugo, who is putting in a floor inside the Red House. There I read him a quote from Hjort's book. "A big, strong iron hook is used, fastened to a slender iron chain, and with a big iron weight attached to the line. Seal blubber is used as bait, and the fish is hauled up with a small, hand-held winch. Using this method, sixty Greenland sharks can be caught in a day."

"Sixty in one day! As if that's anything to brag about," Hugo says with a laugh.

The fishermen that Hjort interviewed were convinced that Greenland sharks wandered far and wide. In April, the boats fished along the coast, but by May they had already moved far from shore. In the summer, the fishermen had to go all the way out to the eastern ice fields of the White Sea off Russia to catch Greenland sharks. And in September many fishing boats headed for the ice floes between Bear Island and Spitsbergen. Seamen who participated in these ventures told Hjort that they often found remnants of nets and hooked lines in the stomachs of Greenland sharks they'd caught in the far north. Back then, that type of equipment wasn't used in the Arctic Ocean, so the sharks must have swallowed them somewhere along Norway's coasts. The fishermen thought the sharks followed the cod on their migration to and from the Arctic Ocean because in the sharks' stomachs they frequently found large quantities of cod that had been swallowed whole.

Toward the end of the section about fishing for Greenland sharks, Hjort makes a general comment that matches exactly what Hugo and I have experienced. "This fishing for Greenland shark is an extraordinarily strenuous business. In these northern waters, storms rage nearly all year round, and then, of course, it's especially unpleasant for the small boats to have to anchor while listing in the cold and heavy swells, hauling up the heavy sharks."[10]

Some of Hjort's informants had wrestled with Greenland sharks all their lives. One arctic seafarer had spent thirty summers in a row chasing them. He said that he alone had probably brought in eighteen thousand gallons of shark livers, which were the only parts of the fish they saved. Thanking him silently, I close up the book about fishing and whaling in northern Norway, written by the good Johan Hjort, who at the time of the writing had a brilliant career ahead of him as a researcher of the vast ocean deep.[11]

The *skrei* are teeming right at our doorstep, and the most agile and hungry of Greenland sharks may have followed them all the way from the Arctic Ocean. Even if we'd had a decent boat, we couldn't have gone out right now, because the big Skrova *skrei*-fishing championship and festival is fast approaching.

26

We had discussed the plans for the event last year after we woke up one morning to discover a gale had arrived overnight from the southwest, blowing straight into the bay. Hugo was worried that the same old fourteen-footer we're now using might not have been tied up securely. Back then there was no floating dock at Aasjord Station, and we had moored the boat in front of the Ellingsen station after returning from a fishing trip.

Hugo's gut feeling turned out to be correct. When we arrived at the other side of the bay, we found the boat filled with water. We bailed for half an hour, turning the boat to face into the storm so it could float securely with the sea at its bow. Since we were already on that side of the bay, we decided to go to the championship *skrei*-fishing festival, which was being held at the Ankas Gjestebud restaurant.

Around the corner from the premises, two adults were sprawled in a snowdrift, thrashing around. Maybe they were trying to make a snow cave. A blues band with a well-known Norwegian actor on vocals was clamoring in the background. "A seagull / you were; to think that you / should end like that! / The others take flight in the storm, / you stand on a skerry and scream."

It was not yet noon. The rugged party tent set up outdoors had room for at least a hundred people, but it was in the process of being evacuated. The wind had just made a respectable attempt to blow the whole thing into the sea, people and all.

Inside, the bar was packed with customers drinking heavily and shouting to one another, as if they were still outside in the storm. The patrons were all adults, and the women were just as aggressive as the men. When I went over to the bar to order some wine, the man next to me began staring. Finally, I felt compelled to return his stare.

"Want to fight?" he asked.

In bewilderment, I asked the man politely if he wouldn't mind waiting until I got drunk. He was probably just joking, though he gave no hint of a smile or any other indication that his comment was meant in jest. From where he was sitting some distance away, Hugo had taken note of the incident. When I got back to our table, he asked me what the man had said. He wasn't surprised. He told me the man was well known for breaking people's arms, so I'd better lie low.

Hugo was then reminded of a story from his childhood. One morning, looking out a window at home, he saw a man come storming out of a tent, barging through an opening he'd cut with his own knife. Two others rushed out of the tent after the first man, who went racing to the shore. Their tent erupted in flames behind them. A fight had broken out and someone knocked over a Primus stove. The man who was being chased reached the sea and began swimming for his life. The two others got out a shotgun and began shooting at the swimmer, who was trying to make it to his boat, which was moored to a float 150 feet out in the waters of Innersundet.

The next day the sheriff appeared on the scene. Someone must have phoned him. The long arm of the law forced the three men to shake hands and ordered them to split the bill for the burned-down tent. All three quickly agreed to the terms, and with that the matter was closed.

Mette joined us at the *skrei*-fishing festival at Ankas Gjestebud. She has strong nerves and always enjoys a festive scene. But the boisterous and slightly barbaric energy of that place, crammed as it was with drunken people who normally don't drink a lot—prompting them to collectively enter a sort of bacchanalian state in which just about anything was permitted—proved too much for Mette. She quickly made her exit.

Hugo and I stayed seated at our table, a bit defiantly because the mood of the place made us nervous. It wasn't immediately obvious whether this was the world championship in *skrei* fishing or in drinking. Everyone's normal reticence had been replaced by an uninhibited, blunt style, which isn't easy to deal with unless you've personally been part of building up the energy level from the start. To cope, we drank a lot of red wine. Fortunately, the festival ended at four in the afternoon, without anyone falling off the dock or getting chased into the water.

I remember the last thing Hugo said as we walked back to Aasjord Station, our shoulders hunched against the snow and wind.

"We're *never* holding that festival at our place. Never in a million years!"

So it's that very same festival that is going to be held at Aasjord Station five days from now, and for which Hugo had built his fish-crate bar. The Ankas Gjestebud restaurant has closed down. Mette and Hugo were then asked to hold the championship festival on their property. Because of its size, it's the most suitable location on the island. When offered this opportunity, they couldn't say no. They've invested a lot of money in the place and desperately need to generate some income. Much more work has to be done on the station before everything is finished, all of which requires huge sums, and the banks also need to be paid. Maybe it's a bit premature to be organizing such a big festival, but that's what they've decided to do.

Three years ago, Aasjord Station was a real eyesore for anyone coming to Skrova. Practically a stain on the whole island's reputation. The disintegrating walls and the wharf were in terrible shape, rotting and on the verge of collapsing into the sea. It was a clear signal to the world that Skrova, like thousands of other small communities along the Norwegian coast, was about to go under. Aasjord Station was not a picturesque ruin like an ancient castle. Instead, it was an uncomfortable reminder of the relentless advance of "progress," which always seems to walk hand in hand with loss of population and general decay. Those who had stayed had their backs against the wall. It was just a matter of time before they would have to admit defeat. The future was definitely not *here*. Well, maybe that description is not entirely true. The point is, that was the impression broadcast loudly from the grand old fishing station rotting on its poles smack in the middle of Skrova.

Now the new Aasjord Station is going to open its doors for the first time. It's meant to be a memorable day, not only for Mette and Hugo, and not just for the station itself, but for all of Skrova. The goal is for Aasjord Station to become a community and cultural center for the whole island—in other words, Skrova's "living room." Over the years the old fishing station has received countless millions of *skrei*. What could be more appropriate than for the first event at Aasjord Station, now that it has risen from the dead, to be a festival celebrating *skrei* fishing?

The last days before the big event are very hectic. Hundreds of guests are expected, more people than live on all of Skrova. Folks will come over from Svolvær and Kabelvåg in big RIBs, fishing boats, and even by helicopter. All the hotels in the region are fully booked, because people travel long distances to take part in the world championship. The goal is not to win but to enjoy the whole setting: the beautiful landscape, the hundreds of boats on the water (weather permitting), and the lengthy *skrei* dinners. Companies from all over Norway bring their employees or business associates to encourage team spirit and enthusiasm. All that said, the parties, and not just the one to be held at Aasjord Station, are the main focus.

Mette and Hugo have already been working practically 24/7 for weeks to handle all the planning, ordering, permits, and thousands of practical matters that need to be in place—everything from extra electrical power to applications for a liquor license and permission from the local fire department. Walls need to be whitewashed, more bars constructed, railings put up, rooms cleared and decorated. They also need to put in a kitchen, because they plan to serve whale meat and fish burgers. They borrow a lot of things from everyone on the island, but other things have to be brought over from Svolvær.

Hugo has even managed to get hold of an old boiler, weighing several tons. It took a crane to deliver it to the wharf and then it was rolled through the double doors into the area once used for putting bait on hooks. Since there is no drivable road to Aasjord Station, anything over a certain weight and size has to be brought by boat. The *Havgull*, formerly owned by the Aasjord family, lowers a pallet with 1,512 cans of beer, plus a tank holding 260 gallons of diesel oil for the heaters, from its deck down to the wharf.

—

It's obvious that all of Skrova wants the celebration to go well. Mette and Hugo notice that the island's most influential residents have taken the event under their wings, secretly pulling strings if the municipal bureaucracy proves to be unnecessarily obstructive. Huge men arrive, dragging heavy items. People I've never seen before—since Hugo usually keeps to himself here on Skrova—are running around everywhere. It's almost as if everyone has an innate understanding of what has to be done. Watching Aasjord Station preparing for the party, with people eager to help practically popping out of the floorboards, makes me think of the Disney film *Cinderella*.

Even the weather is showing its best side, with clear skies and the perfect dry conditions for the *skrei*. The sea is blue and white, like in a beer commercial. By the time the hired musicians arrive by ferry from Bodø on Friday afternoon, almost everything is ready.

27

By ten in the morning small groups of people are already drifting into the rooms. Some of them probably haven't been inside Aasjord Station in forty years, and they're curious to see what it looks like now. All day long steady streams of guests arrive. Several restored old fishing boats, manned by youthful retirees who have hung *skrei* to dry from the masts and rigging, tie up along the newly restored wharf.

Over the years, some of the guests have developed a certain reputation for showing off their brawling skills at parties. Big guys with hands that make a pint glass look like a small water glass. Many drink heavily; some had been hitting the bottle for days. But no fights seem to be brewing. The atmosphere is friendly, if not deferential.

Most of the guests who stay for a long time are from the local area, meaning both sides of Vestfjorden. Hugo spots a man he hasn't seen in fifty years, not since he was on summer vacation, visiting his great-grandmother in Fleines up in Vesterålen. Hugo tells me that when it was hot, the other boy would wear brown long johns under his shorts. And the boy had a pet crow. When they end up standing next to each other, Hugo suddenly turns to him and says, "Do you know someone who used to have a pet crow?"

The man gives a start. He'd almost forgotten about that.

Out on the wharf I happen to fall into conversation with a fisherman from Hamarøy. He does a lot of halibut fishing, and he tells me that it's so frustrating when Greenland sharks get into the nets and tear them to shreds. If Hugo and I don't have any luck on our own, we can go out with him, because there's certain to be plenty of Greenland sharks where he is. I take note of his name but tell him that Hugo and I probably need to do this our own way.

Everyone is eating and drinking. The supply of hard liquor disappears so fast that you'd think this was Great Liquor Day. We have to order more from Svolvær. The most frequently heard comment all afternoon is "The booze is coming on the ferry." When the ferryboat finally enters the bay, many alert pairs of eyes follow its progress toward Aasjord Station. An elderly man wearing a captain's hat orders fifty shots of aquavit with a sardonic grin. He and his small group slowly down every single one, then go out to board their boat. In the meantime, the tide has gone out, so they have to climb down from the wharf. Those guys have done the same thing so many times before that they hardly need to be conscious at all.

As they leave, a sixty-five-foot-long Viking ship, built in the traditional manner, comes sailing into the bay, docking at the Aasjord wharf. It's a newly built vessel with a dragon's head at both ends of the symmetrical hull.

There's a milder, more positive mood spreading through the crowd than last year. Just like before, the mood is self-propagating. The difference is that now it's on an upward swing.

Almost all evening the sky is starry and clear over the Lofoten Sea. As the party winds down, I go for a walk along the wharf. A few snow crystals drift slowly down over the dark rooftops, over the docks and the rocky shoreline of Skrova. The blues music coming from the old saltery makes its way into the most distant crannies of the station; the thump of bass guitar rises up to the attic and sinks down among the wharf posts. The sound hovers over the water in the bay, where the current is constantly moving, and continues out into Vestfjorden.

Skrova is usually completely silent in the evening. With the exception of the wind and a refrigeration unit or a fan that's always going outside the Ellingsen station, there's rarely a sound. The seagulls hardly ever make much noise, since they rarely have to fight for food. Right now the music and laughter inside blend with the airy snowflakes slowly drifting down to melt in the sea. On the bottom, the *skrei* are swimming, waiting to spawn.

The windows of Aasjord Station are glowing faintly, and a lantern hanging from the mast of a boat casts a gentle light over the white façade of the buildings. Is this the first time in history that a dried fish warehouse has been artificially heated? After so many decades of silence and decay, the resurrected station emanates a new energy, like when a new year settles in, driving out the previous one. I think of the whole Aasjord Station as a huge invisible clock that stopped decades ago. Tonight it has started to tick again.

28

It takes us two days to clean up. After that we can focus on our hunt for the Greenland shark. The fourteen-footer is now in considerably better shape, since most of the ice in the double hull has melted, and the boat has been pumped out. But in the morning an icy gale blows in from the east, and Vestfjorden turns white. We can forget about going out. A steady wind creates sharp ice crystals, which glitter in the low winter sun.

We've had a Greenland shark on the hook before, and it will happen again. But not this time. The weather doesn't improve before I have to head back south. During my stay here, the shark hook didn't even touch the sea. But the *skrei* sway in the cold wind. And that is a satisfying, even beautiful, sight.

Spring

29

Spring arrives, and once again my internal compass points north. As the Norwegian author Rolf Jacobsen writes in his much-quoted poem: "Long is this country / Most of it to the north." But when you arrive in the north, most of it is actually south.

Of the four directions, north has always been the one most enveloped in mystery. Until recently, the extreme north was a place that lay beyond the horizon and way out of reach. How it was depicted was limited only by the imagination. The story of the mythic north began early with the prominent Greek astronomer and geographer Pytheas of Massalia. In the fourth century BC, he sailed from the Mediterranean to what is today England. He continued northward along the British Isles to the northern tip of Scotland. From there he set a northerly course for six days, until he came to an unknown land—a place that was completely dark in the winter, while in the summertime the sun shone round the clock. The people were friendly and had a number of peculiar customs. It was foggy, and the sea was frozen over. Pytheas called this land Thule.

Everything Pytheas wrote has since disappeared. Fragments of his annals have survived only because they were mentioned in other works. But people have continued to discuss his travels for more than two thousand years. Where exactly was this place that Pytheas visited? Was it the Orkneys, the Shetland Islands, the Baltic, Iceland, Norway, or maybe Greenland?

In the opinion of the Greek geographer Strabo, the whole story was pure fiction and Pytheas was a charlatan. Everyone knew that the British Isles constituted the northernmost inhabited area in the world. Only Ireland was more barbaric. There men lay with their sisters, and they ate their own parents when they got old. Therefore Pytheas's mysterious land of Thule had to be mere invention.

But the myth of Thule only grew bigger over the centuries. The Roman poet Virgil used the name Ultima Thule—meaning the most distant and farthermost Thule, a shadowy world in the far north. The land on the way toward the night.

The famous Norwegian explorer, scientist, and diplomat Fridtjof Nansen didn't have the slightest doubt. Only one place or region fit all the details we have from Pytheas's description, and it wasn't the Shetlands or Iceland. It had to be northern Norway. Although maybe not everything matched, because the frozen Arctic Ocean that Pytheas described in his account did not fit. But then again, the North Atlantic could have been significantly colder twenty-four hundred years ago. Nansen also suggested that the Norwegians could have told the Greek about the Arctic Ocean, possibly when Pytheas traveled along the coast of Helgeland or even farther north. There he might have experienced the midnight sun. Maybe Thule is the island of Værøy, which Hugo and I can glimpse far out at sea when we take up position near the Skrova lighthouse.

Nansen also wrote about the Hyperboreans. According to Greek mythology, these people lived north of the north wind, near the northernmost sea where the stars went to rest and the moon was so close that you could see all the details on its surface. The Hyperboreans would sometimes invite the god Apollo to attend a dinner and a dance. Some claimed that an enormous temple existed in that land. It was shaped like a sphere that hovered in midair, borne by the winds. The Hyperboreans were also very musical, and they spent most of their days playing the flute and lyre. They knew nothing of war or injustice; they never got old or fell ill. In other words, they were immortal. When they grew weary of life, they would throw themselves off a cliff, with garlands adorning their hair.

Thule, the Hyperboreans, and other mythic stories about the north are not marked by desolation but rather by beauty, purity, silence—and a great longing for all these things. The unknown north was a sort of haven or refuge for something exalted, something we could not exploit, something virginal and pure—as in *innocently virtuous.*

Thule is no longer a dream of somewhere beyond everything else in the world, but it's still a place for which we long.

In mid-May I find myself once again on board the catamaran that will take me from Bodø to Skrova. Cold, mineral-rich water has been stirred up from the depths by ocean currents and winter storms. The sun has given the sea new life, and the ocean flora and plankton are blooming in vast numbers.

Outside Skrova the water is a milky light green. Many seas are named for their characteristic color. The Red Sea probably got its name from the reddish algae. The White Sea is covered with ice most of the year. Storms transport sand particles from the Gobi Desert to the surface of what is called the Yellow Sea. No one can say for sure how the Black Sea got its name, but it originated in Roman times. Maybe the Black Sea is actually blacker than other seas because it contains more freshwater.

Today the waters of the Baltic Sea, the North Sea, and many Norwegian fjords, in particular, are becoming darker as they are being overfertilized by organic matter that absorbs light. A rise in temperature reinforces this development. If the water gets too dark, many of the ecosystems will be damaged or destroyed, but the jellyfish will thrive.[1]

What is the real color of the sea? Over the years, some cantankerous people have tried to dispute the commonly held view, especially among artists, that the sea is blue. They have admitted, almost begrudgingly, that the water does look blue from a distance, under certain circumstances. At least when the sun is shining. Early in the morning the sea is usually a uniform pearly gray. And in the evening, when it's calm, the water reflects the blood-red sunset. Otherwise the color of the ocean changes with the depth, the conditions at the bottom, the salt content, algal growth, pollution, silt from the big rivers, and light from the sky overhead. Various combinations of these factors can lend different hues to the water. The old arctic sea captains knew that the ocean currents from the south bring water that is blue, or bluer than the waters of the Arctic, which is often green.

Right now, the green color of Vestfjorden is caused by the year's first blooming of coccolithophores, a type of flagellate, which are single-celled, chalk-forming plankton propelled forward by a whiplike tail or *flagella*. They can be found by the thousands in every drop of water. Under a microscope, the body of flagellates look like round pebbles with filigree patterns and structures. Normally these types of algae don't appear in such abundance before later in the year, but the sea around us is changing.

Just as most animals on dry land feed on grasses and other

plants, most organisms in the ocean live on plankton. Plankton does the same thing that plants do on land, that is, bind enormous amounts of carbon and produce oxygen through photosynthesis. A particular type of blue-green algae is so productive and abundant that scientists have estimated that this organism alone produces 20 percent of the oxygen on earth. Its existence wasn't even known to scientists until the 1990s. Plankton plays a major role in making the earth habitable. We owe an immeasurable debt to something we can't see—something that most people know almost nothing about.

Plankton can assume the strangest forms. If you looked at photographs taken with an electron microscope, you would hardly believe your eyes. The plankton may look like snow crystals, moon landers, organ pipes, the Eiffel Tower, the Statue of Liberty, communication satellites, fireworks, images from a kaleidoscope, toothbrushes, empty grocery baskets, open waffle irons, wineglasses with ice cubes floating inside, champagne glasses lined with leopard skins, Grecian urns, Etruscan sculptures, bicycle racks, long-handled landing nets, carburetors, feathers, flowers, slime balls with apples inside, Bluetooth headphones, disco balls, melting church bells that are also transparent, flying carpets, lion's teeth, fishing nets, top hats, vacuum cleaners, spermatozoa, brains, and fountain pens. Plankton can take the shape of almost anything that exists in the world, as well as so many unfamiliar forms that you could build a whole other world. Millions of these microorganisms can live in a bucket of clear, clean salt water—including a large number of flagellates, chalk-forming plankton covered with plates called coccoliths.

A billion years ago, choanoflagellates formed colonies and

were possibly the origin of the first multicelled organisms.[2] In that case, they are the ancestors of everything alive today. All of our ancestors by definition managed to propagate in one continuous chain over billions of years, ever since life arose in the sea. This may sound improbable, but that's how it is. We just don't usually view things from this perspective. And why should we?

Evolution is blind and runs like a river through time. It cares nothing for the losers that disappear along the way.

The sea has many colors. But what is the sound of the sea itself? Waves that trickle over a beach or pound against the cliffs and rocks of the weather-beaten coasts? Yes, that's what the ocean sounds like from land. Underwater it's a different story. There the sea has a unique sound, a deep humming that emanates from itself—the moaning of the Behemoths in heat.

For decades, people all over the world have discussed this sound, which only some can hear. It has been described as a diesel engine heard from a long distance; a trembling, low-frequency tone. Some people, even sensible Welshmen, have actually claimed the sound can cause nosebleeds, headaches, and insomnia. Many have tried to explain the phenomenon, theorizing it might be caused by everything from telephone poles, cables, submarines, communication equipment, tinnitus, and mating fish to UFOs. So many perfectly clear-headed people have insisted they can hear the sound that research has been done about it. Now French scientists at the Centre National de la Recherche Scientifique think they may have found the answer.[3] Long sea swells create microscopic activity on the seafloor. Under certain conditions, long, heavy waves cause the ground to tremble, and the vibrations create deep sound waves, which certain people can hear loud and clear.

By the time the ferry arrives at Skrova from Bodø, it's late at night, as usual. But after the darkness of winter the light has made a strong comeback, and for the next two months the sun will hardly set over Aasjord Station. The fall and winter turned out to be rather problematic seasons for two men hunting for a Greenland shark in a small boat. Now, in our fourth season, we're determined to succeed.

As usual, Hugo has made good use of his time. He has done a lot of work on the Red House and installed two bathrooms in the main buildings, to be used for future events held there. He and Mette have also brought over their two Shetland ponies, Luna and Veslegloppa, from Steigen. Now the horses are grazing in a small green valley, several hundred yards in the direction of Hattvika. Hugo is going to clear out the premises of the cod-liver-oil mill, which is now filled with old oak barrels at the back of Aasjord Station, so he can turn the space into a stable for the winter. I've been wondering why they still keep these ponies, now that the children have moved away from home. But that's not the way Hugo and Mette think. They would find it strange if I asked them.

Hugo has been out to have a look at the fin whale that washed ashore on the island of Gimsøy. He tosses two whale baleen onto the table. They're lightweight and look as if they could be made of thin fiberglass. These long, stiff hairs, which hang inside the top of the whale's mouth, catch krill and plankton when the seawater is filtered out. But the baleen are mere trivialities. Hugo wants to have the whale's skull. He's not sure how to make that happen, but he thinks he'll need a freighter.

Upstairs, Hugo shows me several pieces he's working on, using pencil on acid-free cardboard with glued-on pieces of recycled cotton paper from India. The paper has great texture, creating fine nuances in gray and black. Some recognizable objects are visible in the work, including zeppelins, which also look like floating whales. In another piece, what looks like a Greenland shark is depicted turning around in the water.

Hugo is also working on a painting that has the Steigen menhir, or standing stone, as its theme. The stone was northern Norway's tallest menhir and had stood on Engeløya, several miles from Mette and Hugo's house, for fifteen hundred years. Until the day a municipal edge trimmer came along and toppled the stone, breaking its base. Apparently it's beyond repair, but Hugo thinks it can be fixed.

At dinner, which is a small fried halibut that Hugo caught with a pole in Steigen, he shows me a marvel of technological innovation. Mette has presented him with a gift of a deep-sea fishing pole and a powerful Japanese reel with gears. That's what we're going to try out now. I've brought a vest with belts and straps, the type of equipment that deep-sea fishermen use off Bermuda when they're going to reel in sailfish and swordfish.

The line we've used so far—eleven hundred feet in length—weighs a lot and will only fit inside a deep tub if we carefully coil it up by hand. Now we're going to attempt to pull up a Greenland shark weighing as much as two thousand pounds, using a line that's hardly thicker than sewing thread. This is new technology, which supposedly has some of the same characteristics as a spider's web. That may not sound reassuring. But, trust me, it is.

30

Next morning at the crack of dawn, a thick gray haze has settled over both land and sea. Aasjord Station is wrapped in near total silence. All sounds are muted by the fog, but whatever we do hear grabs our attention with extra clarity. Our sense of hearing becomes as acute as our sense of smell.

The sea looks as if it's paralyzed beneath the blanket of fog, absorbing not only sounds but also silence. I hear a fan or a generator I haven't noticed before, the sound coming from somewhere on the other side of the bay.

Three hours later the haze has lifted. Nimbostratus clouds hover in a low gray layer, glowing with a sickly yellow color. The sun will soon break through, so we get everything ready and race across a calm sea, out past the Skrova lighthouse and Flæsa. This time our bait is a real delicacy. Not since the Highland bull have we had anything of such high quality. The bucket of *skrei* livers from the winter has done its thing. Several gallons of pure oil have now formed on top. Hugo will use the oil to make paint. At the bottom of the bucket is the gleaming, stinking brown sludge called *graks*. The *graks* is almost pure fat, and it was this substance that fishermen, including Hugo's grandfather, used in the olden days when they were hunting Greenland sharks. It has a rank smell, but the stench is more complex than the smell of the Highland bull carcass, which reeked only of death. We fill a paint can with *graks*. The smell will be our underwater siren song.

Again we've triangulated our position, using fixed points on land. We've also brought along a GPS, but it has so many settings that neither of us feels entirely confident about using it. Then I toss out the paint can. We've punched a lot of holes in the lid, which is fastened only with a rope so the contents will quickly seep out onto the seafloor. And reach the Greenland shark, where it's waiting.

Is it possible to imagine the shark's world? To picture what it feels like to be surrounded by water and darkness? The shark hardly notices, since that's all it knows. Just as we don't really register the air around our body; it's something we take for granted. The dim, cold deep is the shark's world. Down there it glides around, slowly and soundlessly, like a machine made of muscle. With poison in its blubber, blood, and liver. With lifeless blind eyes from which hang parasites, long larvae that pierce the shark's eyeballs. All the shark wants is to maintain and continue its own existence. It's unlikely to feel anything resembling joy or sorrow, maybe not even pain. Each time the shark devours a seal, or buries its snout in the rotting carcass of a whale, it must register some sort of automatic contentment, knowing that its existence is secured for another month or so. And that's what the shark wants in the world; that's its mission in life: to keep going until the next meal. The only living creatures the shark has contact with are the ones it eats, aside from the time when the female's eggs are fertilized—a process that occurs without any recognizable sign of joy or tenderness. The offspring quickly develop big teeth and begin life as cannibalistic predators in the womb, where the strongest one devours its siblings and then enters the world alone.

When the small Greenland sharks are born, they're able to glimpse a pale gray hue many hundreds of feet above, though they pay little attention to it. Then they start searching for something to eat in the black silence of the lonely cold. The shark does not ask why it happens to exist. All life is programmed with a will to survive. No animal commits suicide, no matter how bleak its Hades-like environs may be.

So that's the inept attempt of a human being to put himself in the shark's world. It may seem bleak and hopeless to us, but maybe the Greenland shark hears an entirely different music rushing through its blood vessels. It's weightless, has no enemies, and floats in a universe to which it has become superbly adapted over tens of millions of years.

No, it's impossible for us to imagine how it feels for a shark to be in the world.

You know the drill. We've tossed out the chum bait, but the actual fishing won't start until the following day.

Hugo switches off the motor, and we let the boat drift. We keep drifting as we chat occasionally, or just sit in silence. The silence is never oppressive for the two of us when we're together, and maybe that's as good a definition of friendship as any other.

After only half an hour, we've drifted so far out that I think I can discern the end of the Lofoten archipelago. Beyond Lofoten Point is the Moskstraumen, the phenomenon whose very name has scared the shit out of seamen for thousands of years.

For millennia the place has been viewed as the sea's navel, the world's well, the bottomless gullet, or as the entrance to the cosmic void of Ginnungagap in Norse mythology. According to some leading medieval minds, Moskstraumen was the point from which the ocean is sucked in and then spewed out in torrents. Maybe the water pops up somewhere else in the world after having passed through the earth's underground circuits. Does the earth suck in the sea whenever it's in need of nourishment? That's what sharp minds claimed centuries ago. Could that be how the tides were regulated? By the water going in and out of the interior of the earth, from Moskstraumen—the place where all the winds meet and create chaos, or where the currents are so strong that they extinguish the winds?

Olaus Magnus called the Moskstraumen the *"Horrenda Caribdis,"* which will suck down anything that gets too close, crushing and swallowing ships, people, and animals. The Norwegian cleric and historian Jonas Rasmus (1649–1718), who was a native of Møre, believed that Odysseus himself actually came to Lofoten and encountered the Moskstraumen. Rasmus reported that the most terrifying and thundering waterfall could be heard between the cliffs; the whirlpools were so huge and powerful that any ship happening into them would be dragged to the bottom.[4] In 1591, the Danish-born bailiff Erik Hansen Schønnebøl described the Moskstraumen as so turbulent and the roaring so mighty "that the Land and the Earth tremble, the houses shake." On a map created in Hamburg in 1683, the "Moskoe-Strohm" is depicted as a calamitous area extending hundreds of sea miles. Author Edgar Allan Poe went even further in his story "A Descent into the Maelström," published in 1841. The story describes how a boat full of local fishermen is sucked down into the whirlpools, which roar louder than Niagara Falls and make the mountains shake.[5] Even Captain Nemo's submarine, the technological wonder called the *Nautilus,* could not come to grips with "a vortex from which no ship has ever been able to escape," where the churning vortex sends "not only ships but whales, and even polar bears" down to a certain death.[6]

31

Since the last time Hugo and I met, I've been in touch with one of the world's foremost Greenland shark researchers, which may not say much, since hardly anyone can really lay claim to that title. His name is Christian Lydersen, and he works at the Norwegian Polar Institute. He has studied various aspects of the life cycle and biology of Greenland sharks. Hugo is interested in hearing what I learned, so I tell him everything I can remember, feeling like a conscientious diplomat delivering a report after a trip to a distant and troubled corner of the world.

Lydersen and the other scientists have done fieldwork off the west coast of Svalbard. After talking to experienced hunters, they set out a line with twenty-eight shark hooks from the research ship *Lance.* They used ordinary halibut line made of nylon, with wire as the leader, and they baited the hooks with blubber from a bearded seal. The line was lowered into declivities down to a depth of a thousand feet.

On the very first attempt, they had Greenland sharks snagged on every third hook. Soon they had caught forty-five sharks, more than they needed to find out what they wanted to know about the effects of diet, genetics, and pollution. Some of the sharks that were pulled up had only the head remaining. While they had hung defenseless on the line, their fellow sharks had eaten their entire body. In the stomachs of those that were still intact when they were hauled on board the ship, the scientists found ring seal, bearded seal, hooded seal, and the remains of minke whales, as well as cod, wolffish, haddock, and other types of fish. The Greenland sharks had swallowed whole cod weighing more than eight pounds, along with wolffish that weighed twice that.

It's out of the question that Greenland sharks are able to kill whales, but Lydersen found the answer to where the minke whale blubber had come from. Genetic samples are taken from every minke whale caught by a Norwegian boat. There's no market for the blubber, so it's tossed overboard. And guess who gorges on the stuff on the seafloor?

So how does the Greenland shark catch seals? Lydersen and his colleagues discovered the answer to something Hugo already knew. The seal flesh couldn't be merely carrion, because far too much of the seals have ended up in the sharks' bellies. The seals must have been taken alive. But how? The scientists fastened sensors to some of the sharks and then released them. The measurements revealed that the sharks actually swim slower than seals and all other fish. There was nothing to indicate that they have the capacity to propel themselves in small, brief lurches. So they couldn't possibly catch species that are much faster simply by using ordinary hunting techniques. The answer lies in the fact that ring seals, harbor seals, bearded seals, and hooded seals are highly evolved mammals. This gives them many advantages, but also one major weakness: they sleep much like we do, slumbering deeply, with their eyes closed, and with both hemispheres of their brain shut down (so-called bilateral symmetrical sleep).[7] The seals lie on the seafloor, dreaming—maybe about shoals of fish, about mating and playing, about relatives, or . . . Well, it would be interesting to know exactly what seals do dream about. Those that sleep on top of the ice or on the surface of the water can sink into such deep REM sleep that you can almost pull up alongside them in a motorboat before they react. On the ice, polar bears are a constant threat. Maybe the seals feel safer on the seafloor, where they may sleep more lightly or for briefer periods. But not even there are they safe. A dark, cigar-shaped shadow glides slowly and silently along the bottom in search of food. Patiently and deliberately, with its ampullae like a sort of electromagnetic radar able to detect life. A sleeping seal must be easy prey.

The Greenland shark takes its time, attacking with its double rows of saw-edged teeth. By the time the seal abruptly awakens, it is already locked into the shark's stinking jaw and about to be gnawed to death. Maybe the seal is paralyzed by shock and terror, having been yanked out of its dream world to experience this last, brief nightmare of its life. That makes me think of something the German filmmaker Werner Herzog once wrote: "Life in the oceans must be sheer hell. A vast, merciless hell of permanent and immediate danger. So much of hell that during evolution some species—including man—crawled, fled onto some small continents of solid land, where the Lessons of Darkness continue."[8]

"Holy shit," says Hugo, adding that you'd have to be pretty depressed to think of the ocean that way.

"But how do Greenland sharks catch fish?" I ask rhetorically.

By attaching advanced transmitters to the Greenland sharks, Lydersen and his colleagues learned a lot about their wanderings. The transmitters were attached in the western Svalbard area and were programmed to fall off after a maximum of two hundred days. Some showed up near Greenland, others in Russian waters in the south Barents Sea. Many were never found, presumably because the sharks who had them on their backs were under the ice when the transmitters came off. One shark had wandered six hundred miles in fifty-nine days—a surprising distance considering how slowly the Greenland shark swims. They stayed largely in relatively shallow water, between 150 and 650 feet deep. But one shark went as deep as the instruments were capable of measuring, which was 5,118 feet. And it probably went even deeper. Lydersen and the other scientists also discovered that some sharks may move between the Atlantic Ocean and the Pacific via the Bering Strait.

No matter what, tests of the Greenland sharks' livers and blubber proved that the worst, most persistent environmental toxins circulating in the ecosystems accumulate in the northern regions, all the way up to the North Pole, and end up inside the polar animals, including Greenland sharks. Some of the poisons cause species to change gender; others can destroy their ability to propagate and lead to cancers and other diseases. The Greenland shark carries an even higher level of poison than the polar bear—and a dead polar bear is considered toxic waste.

Like so many times before, we drift across Vestfjorden in our rubber boat, floating above the seafloor's unseen landscape of forests, valleys, mountains, crags, deserts, and plains. It's a clear and calm day, with small ripples shimmering like fish scales.

Usually we're all alone when we drift like this, bobbing in the water. Occasionally we might see a modern little plastic boat fishing in the area. If it's clear weather, we can see freighters with illuminated wheelhouses gliding soundlessly in or out of Vestfjorden, heading for Narvik along the shipping lane located six miles away. We never see any pleasure boats, but now an RIB is coming straight toward us. It's hydroplaning and approaching as if aiming for us. Hugo and I exchange glances. The situation makes us think of the time we were on our way out of Flaggsundet between Engeløya and the mainland in Steigen.

It was a clear summer night with calm seas, bright with the midnight sun. We didn't see any other boats on the water, so Hugo gave the good old fourteen-footer full throttle and headed for the fishing spot we had chosen. I sat in the bow, blocking Hugo's view forward as the boat hydroplaned. But there was nothing to see except smooth water. We'd made sure of that when we started out. I sat facing Hugo, looking away from the direction in which we were traveling. After ten minutes or so, I saw his face contort as he suddenly twisted his whole torso ninety degrees and yanked the tiller of the outboard motor so the boat abruptly veered to port, while gravity threw me toward starboard. I barely managed to hold on. A hundredth of a second later, because this was the sort of situation that happens in slow motion, I found myself staring straight into the terrified faces of two men. They were so close I could have shaken hands with them. As we brushed past their little boat, both men were standing up, and when the waves struck the side of their vessel, they were in danger of falling overboard.

The two men had set out for the same reason we had. Ostensibly to go fishing, but mostly just to be out on the water on such a perfect summer night. For ten minutes they'd watched us coming closer. They must have grown increasingly anxious, exchanging glances and asking each other when we were going to change course. Maybe they assured each other that we must have seen them. The alternative was unthinkable.

If our small plastic boat had struck their small plastic boat, in the middle of the fjord, with perfect visibility and without even a gust of wind, it would have been the most idiotic accident to occur along these coasts in decades. All four of us could have been mown down, and the investigators would have had to consider whether we had deliberately struck the other boat.

After we had a good laugh over it, I asked Hugo, "What's the likelihood that two boats would run into each other like this by accident? Zero chance, right?"

"You've got it ass backward," he replied. "They were in the middle of the shipping lane, which is narrow and flanked by shallows. Since we failed to see them right off the bat, the chance of us running into them was not negligible. It was sky-high."

When we were only a few yards away, Hugo had suddenly caught sight of two men running back and forth in panic on either side of my head, like in some frenzied puppet show. Then they stopped moving, and one of them tried to start their outboard motor.

The next day we ran into those same guys at a concert at Steigarheim. One of them came over to Hugo, looking angry, and asked him what the hell we were doing. They'd been just about to jump into the water. And they weren't wearing life vests. "We were," Hugo told him, adding coldly that everyone is legally required to wear life vests out at sea.

Now the RIB approaching us at great speed not far from

the Skrova lighthouse veers off in plenty of time and continues around the island.

As usual, the strong currents carry us far away. Hugo starts the motor, and we head closer to land to catch some fish for dinner. As we go, he teaches me a few new words. He points toward shore, where a promontory sticks out in our direction and continues underwater, far out into the sea. Hugo says that the Norwegian word for this type of promontory is a *snag*. Many fishermen still have a rich vocabulary to describe various sea-floor conditions—or, for that matter, significant nuances in the halo around the moon.

The landscapes on shore naturally continue underwater. If we drained the ocean, this would be much more evident. But where would we put all that water? I happen to think of a story from ancient Greece. If I remember correctly, an old king had made a wager, and if he lost, he would have to empty the sea of all its water. After a while the winner of the bet came to visit and asked when the king was planning to start emptying the ocean. The king replied that he was just waiting for the lucky winner to stop the water from running into the sea from all the rivers and streams, since that particular task had not been part of the bargain.

There's a lot of fish along the sides of the *snag*, and after a few minutes we've caught two kelp cods for dinner. Like the *skrei*, they are in the cod family, but the kelp cod are stationary. They have a deep red color, making them very hard to spot by preda-tors in the red, yellow, and brown kelp forests.

On a day like today, Vestfjorden may seem like a paradise of purity. That's far from the truth. Even though these are open waters with such strong currents that little debris is left lying around, we do see discarded plastic items floating in the sea. Maybe from local communities, maybe from some distant coast. The world ocean is an interconnected element.

Twenty years ago a container ship on its way from China to the United States ran into a winter storm in the Pacific Ocean. Some containers came loose, broke open, and landed in the sea. Since then 28,800 plastic bath toys—blue turtles, green frogs, and yellow ducks—have been carried all around the globe by the ocean currents. One writer followed the yellow plastic ducks around the world and back to the factory in China where they were made. He titled his book *Moby-Duck*.[9]

Like all other types of plastic, the ducks don't sink. At least not until they dissolve into microscopic particles. Plastic, and many of the toxins it contains, won't break down for thousands of years. Some comes from the rinse water in washing machines when we wash synthetic fabrics. Because of the ocean currents, gigantic islands of spinning plastic collect at specific locations, where they whirl around in spirals. One such maelstromlike island of plastic in the Pacific Ocean is reportedly half the size of Texas. Another is piling up in the north, in the Barents Sea. There, even the crabs have plastic in their stomachs. When the plastic disintegrates into microparticles, it's ingested by plankton, or it sinks to the seafloor, where the bottom animals eat it.

So this is not a cute story about yellow rubber duckies bobbing around in the big bathtub of the ocean. When scientists examine Norwegian seabirds, they find that nine out of ten have plastic in their stomachs. The birds can't digest the plastic, and it prevents them from taking in nutrients. Every year more than a million seabirds die, as well as more than a hundred thousand marine mammals, because of plastic garbage.

Cod, which swim around with their mouths open, can also end up with their stomachs full of plastic. In the Mediterranean, young sperm whales sometimes wash ashore, and the cause of death often remains a mystery. But when one such sperm whale was examined, they found thirty-seven pounds of nondegradable plastic in its stomach. The most likely cause of death was big sheets of heavy plastic from the numerous greenhouses in southern Spain.[10]

Here in Norway, we too give the sea a good beating. In the fjords, the fish farms are allowed to release as many toxins as they like. Trawlers drag steel dredges across the seafloor, leaving behind a desert. Until recently, we thought coral reefs existed only in the tropics, in relatively shallow water. But there are countless cold-water reefs off the Norwegian coast.

Way off the coast of Lofoten, near Røst, is the largest deep-sea reef ever discovered anywhere, so far. It's close to twenty-five miles long and two miles wide, situated in rugged terrain at a depth of more than one thousand feet near Eggakanten, where the continental shelf ends. The Greenland shark is by far the longest living vertebrate, but no large organisms on earth live longer than corals. The ones near Røst (which belong to the genus *Lophelia*) may be eighty-five hundred years old—considerably older than the age assigned to the earth itself only a century ago. Fishermen have always known that coral reefs are teeming with life. Large numbers of fish and bottom animals find food and protection in the forests of coral, among the red or pink branches of bubblegum coral (*Paragorgia arborea*), which can grow to be sixteen feet tall. But when a trawler drags a steel dredge along the bottom, the coral is destroyed in a matter of seconds. Full trawling nets are hauled up from the reefs, but the method can be used in that spot only once.

The colorful spawning grounds are literally as fragile as porcelain. When the coral reefs are shattered, it takes several thousand years for them to be restored to the same size. It would be hard to think of anything that is more shortsighted. It's like sawing down trees in orchards to harvest the fruit or nuts.

It's true that today some of the big reefs off Norway are protected areas. But many have not yet been mapped, and new deep-sea reefs are regularly being discovered off the Norwegian coast and in the Barents Sea. By then they're often badly damaged by trawlers, and the broken skeletons of coral forests are strewn all around. Oil companies have received, and will continue to receive, permission to drill for oil on and near protected Norwegian coral reefs.

The machine churns on. Many places are now being opened for kelp trawling, also off Skrova. This is happening despite recommendations from scientists and protests from coastal fishermen. Small fish spawn in kelp forests, and large numbers of species, including the kind of kelp cod we just caught, also live there. Nevertheless, the authorities are allowing this important and vulnerable ecosystem to be destroyed because someone wants to make money by selling kelp.[11] The kelp is yanked up with big grapple buckets. It has become a billion-kroner industry. One boat can harvest up to three hundred tons of kelp a day.

—

Who wants to think about things like that after a perfect day on Vestfjorden? Not Hugo or me. After eating the kelp cod, we sit down against the sunny wall. Huge super-RIBs from Henningsvær, Kabelvåg, and Svolvær are passing in front of Aasjord Station at regular intervals, filled with sightseeing tourists.

The tourists come here because of the landscape, which is considered uniquely beautiful. People from other parts of the world pay big money to see the splendor with their own eyes. I can understand why. I think of the dramatic peaks sticking straight up from the sea, the eternally changing light in both summer and winter, white sandy shores, pale green grass on a narrow hat brim of land with a backdrop of vertical mountains and small glaciers, the ocean's great wealth of life, and an old and relatively intact cultural landscape as well. Oh yes, Lofoten has so much to offer that I can see why one international travel magazine after another has described our island realm as perhaps the most beautiful in the whole wide world.

But such an appraisal is not a given. Our view of what is beautiful isn't timeless. This becomes eminently clear when we read older descriptions of Lofoten.

In 1827, the Norwegian Gustav Peter Blom—district magistrate, member of the first national assembly at Eidsvoll, and later county administrator in Buskerud—embarked on a journey through northern Norway. When he returned, he described his impressions and experiences in his book *Bemærkninger paa en reise i nordlandene og igjennem Lapland til Stockholm i aaret 1827* (Remarks on a Journey in the Northern Lands and Through Lapland to Stockholm in the Year 1827). Blom's view of nature in Lofoten was not only tepid, it was bluntly dismissive. In his opinion, the Helgeland coast might be downright ugly, but Lofoten took the cake. There it wasn't possible even to *imagine* any natural beauty, except in a yearning for the same. Blom writes: "Lofoten is as devoid of natural beauty as is possible. The steep, high cliffs plunge right down to the sea and only rarely allow space for a solitary house . . . That any of these places might be considered beautiful is out of the question, but the ugliest is undeniably the strait area in Flakstad parish.— It's situated on a bare cliff near a narrow harbor, closed off by skerries and islands, providing little ground for buildings; and above looms a steep mountain wall, threatening to crash down upon both the houses and the harbor."[12]

In places where I frequently see views of dazzling beauty, Blom saw only an eerie, barren, and desolate landscape totally lacking in appeal. Blom writes that the east coast of Lofoten, which is where Hugo and I are spending our time, is ugly. But for him, nothing surpasses the vulgarity of Lofoten's western side. There treacherous winds blow, and nature is particularly hideous.

Blom most likely visited Skrova, since he mentions both Vågakallen, the highest peak in Lofoten, and the town of Brettesnes on Storemolla. Skrova is approximately halfway in between. Whenever we're close to the Skrova lighthouse, we can see Vågakallen (3,091 feet), unless there is fog or snow. Blom writes that the mountain resembles an "old fisherman wearing a cap and carrying his sail under his arm, and hence its name." (In Norwegian, *kallen* means an old geezer.) In the opposite direction, toward the northeast, are Lillemolla and Storemolla, where the mountains are only half as high, but closer to us, so their presence is more noticeable.

Unlike Blom, Kaiser Wilhelm II of Germany was obsessed with the natural beauty of the Norwegian fjords and coastline—and especially Lofoten. Accompanied by an entourage of yachts and other marine vessels, he insisted on taking in the famed purple colors of the north, "the sea's floating gold, unmatched by either the Alps or the tropics, either Egypt or the Andes Mountains."[13]

Kaiser Wilhelm decided to visit Lofoten after seeing a painting in Berlin in 1888. Using photographs as the basis for his work, the exhibitor had created a panoramic painting, 377 feet wide. The photos had been taken from the village of Digermulen, right behind Storemolla. If the pictures were taken today, there's a chance our little boat might end up in the scene.

The Kaiser's favorite painter was the Norwegian Eilert Adelsteen Normann (1848–1918), who worked in Berlin and painted Lofoten. Unlike Christian Krohg, Normann managed to depict Lofoten in all its glory. He even succeeded in portraying the "floating gold" of the midnight sun, and without going mad from observing the flooding light, as Lars Hertervig had claimed happened to him. For an artist who wishes to paint Lofoten, it's an advantage to have grown up in the area. All the famous realistic Lofoten painters from the late nineteenth century and early twentieth century were from the region. Normann was from Vågøya, near the southern entrance to Vestfjorden. Gunnar Berg (1863–1893) was from Svinøya, in Svolvær, as was Halfdan Hauge (1892–1976). Ole Juul (1852–1927) was from Dypfjord, near Henningsvær, and Einar Berger (1890–1961) was from Reinøya, in Troms.

As a boy, Hugo sometimes threw snowballs at the window of Halfdan Hauge's studio on Svinøya. He remembers the artist as an elegant old man. Normann was the cousin of Hugo's great-grandfather.

Hugo is an abstract painter, but he has a deep respect for tradition.

The Lofoten Wall is like rows of black shark teeth, one behind the other. For hundreds of millions of years, the sea has slammed against this barrier, to little avail. Even the ocean fails to make an impact when confronted by the Lofoten Wall. From a distance it may look like an impenetrable fortress of stone, and in many ways that's exactly what it is.

Parts of the mountains that make up the wall are about three billion years old. Not the Lofoten Wall itself, but the rocks forming the peaks.

As usual, I've brought along a small stack of books, this time several volumes about geology and the early history of the earth. When Hugo goes out to continue his carpentry work at the Red House, where the electricians and plumbers will soon be able to come in and finish the job, I stay behind to read.

One book is about the age of the earth, or rather about the history of our ideas on that topic. In 1650, the Irish bishop James Ussher calculated that God had created the world on Saturday, October 22, in the year 4004 BC—at approximately six o'clock in the evening. Ussher was widely read and admired. He based his theory on the chronology in the Bible, just as people had done before him and continued to do afterward. Today, such a notion may induce laughter, but in the bishop's day no one imagined that the earth had existed before we did.

Over the next centuries, there were numerous indications that this calculation was completely crazy. Fossils from enormous sea animals were found far from the ocean, sometimes on mountaintops and in the very clay beneath Paris, which had apparently been underwater long, long ago. What had happened to all these strange creatures? Many species had evidently been extinct for eons.

Some sharp minds, such as the English astronomer Edmond Halley (the man behind Halley's comet), attempted to establish the age of the earth by calculating the amount of salt the rivers carry out to sea. For the ocean to have become as salty as it is, the earth had to be considerably older than a few thousand years.

In the eighteenth century, philosophers and naturalists began to realize that the earth must be at least tens of thousands of years old. Many kept this notion to themselves, fearing the wrath of the church, but it was clear that Ussher's calculations were seriously misleading. As geology slowly but surely became established as a science, more people understood that the earth had to be much, much older than the Bible claimed, maybe even by millions of years. Sedimentation, eroded mountains, and studies of volcanoes left little doubt. Eons ago, North America had been tropical, and India was covered with ice. Apparently most of the earth had been underwater at one time or another. This was difficult to deny, but how should it be interpreted? Couldn't the discovery of shells and fish fossils on mountaintops be proof that the Deluge had actually taken place, albeit much further back in time than initially thought? Or was it proof that God could eradicate species that did not please Him?

It became fashionable to collect fossils, even among enthusiastic amateurs. Some of the discoveries came from extinct species like mammoths, dinosaurs, and giant reptiles from the sea. Or trilobites and ammonites, a subclass of Cephalopoda (today's octopuses and squids) that had shells resembling coiled rams' horns and existed in thirty to forty thousand different types before they went extinct. Especially all the strange teeth that were found caused concern. Some looked like shark teeth, but they were much too big. Was it possible that giant sharks and other prehistoric creatures could still be found in the ocean deep, as *living* fossils?

For a long time the question of the earth's age continued to be a philosophical and theological issue. But in the nineteenth century more people became aware that the earth was, in fact, infinitely older than everyone had believed. This meant that almost the entire history of the earth had been played out without *us* being present. This was not something that was easy to accept, since it represented a radical break with the religious worldview. The earth couldn't possibly have been created a few thousand years ago over the course of six days. Suddenly it looked as if human beings had entered the scene recently, after other species had existed for many millions, maybe even billions, of years.[14]

We're used to viewing the earth's geography and the location of the continents as static. But from a geological perspective, that is far from the truth, and Lofoten is one of many examples offering proof. A billion years ago, the landmasses that would become Scandinavia were located close to what was the South

Pole. Or, to be more precise: Scandinavia was located where the South Pole then existed, because the poles have also moved around, even traded places with each other.

Scandinavia was part of the primordial continent Rodinia, which, after several hundred million years, broke up into many smaller continents. One of them is today called Baltica. Over the course of several million years it joined with Laurentia (North America and Greenland) to form the temporary super-continent of Euramerica. When these two continents drifted toward each other and collided, mountain chains were created on either side. Laurentia and Baltica drifted apart again, and in the process a new ocean was formed. This happened not once but twice.

And so it continued. Three hundred million years ago the landmasses on earth gathered into one connected continent called Pangaea. Two hundred million years later, Pangaea also broke up into pieces. At the end of the sixteenth century, the Flemish cartographer and geographer Abraham Ortelius noticed something striking. If you move the east coast of South America toward the west coast of Africa, the two parts fit together like pieces of a puzzle. But as late as 1912, when the German polar researcher and geophysicist Alfred Wegener published a groundbreaking work on continental drift, the Pangaea theory was still considered highly controversial.

Molten rock from the earth's core spewed up, hardening over the primordial seas to create new land. The continents drifted around on the earth's crust like unmoored ships. The Ice Age pressed them together like the first floor of a collapsed high-rise. The earth's plates broke apart, smashed together, traded places, and wandered on, frequently with large pieces of other continents attached, as collateral damage.

Vestfjorden is not a classic fjord. It's a sediment basin. Beneath us there are many miles of soft, sedimentary rock.[15] During the last Ice Age, when the ice sheet was miles thick and covered more of the Scandinavian Peninsula, some peaks of the Lofoten Wall still stuck up above the ice cap as what's called *nunatak,* or glacial islands. In fact, the wall was responsible for the ice being steered southward.

The Lofoten Wall is partly composed of the oldest and hardest types of rock on earth. They were formed at the same time as the first single-celled animals arose in the sea. But other parts of the Lofoten Wall consist of much younger remains from the collision between Laurentia and Baltica. Over the course of millions of years, the continents were forced toward each other, rather like elevator doors, except that they didn't stop when they encountered resistance. Instead, they crushed each other, so that the mountain masses rose up and were shoved from one continent to the other.

That was how the mountain chains like the Himalayas, the Andes, the Rockies, the Alps, and the jagged coast along Lofoten, Vesterålen, and Senja were formed.

By the way, the work of the Roman author and naturalist Pliny the Elder (AD 23–79) is the first known source in which our part of the world is called Scandinavia (*Scadinauia*). It means a torn up, dangerous, or damaged coast. It was the damage from the big glaciers, which gnawed the land to pieces and made it what it is, with fjords, islets, and archipelagos. And hardly any place is more beautiful than Lofoten—depending, of course, on the eye of the beholder.

Even the Lofoten Wall is not eternal or immutable. And yet it may well be the closest we'll ever come to that.

32

The evening is so lovely that we decide to go out on Vestfjorden. The mountains are reflected in the water, something that hasn't been seen in months, according to Hugo. He claims the area always has fantastic weather whenever I come north. It's not true, of course, but I tell him I've got connections among the descendants of the wind merchants, and some of them are still in the business of conjuring the weather. Hugo laughs.

"You don't believe me? They told me the magic works even if you don't believe in it, so what you think doesn't make the slightest difference," I say.

We keep our voices low as we talk, as if the fish might be listening. That's how quiet it's been all day. But now we can see that something is happening in the west. Restless and agitated movement is almost always noticeable out there among the sky, clouds, wind, and sea—a permanent drama that we always observe from a distance, because when we're in the midst of it all, visibility drops to nil.

Shadows slide across the gray, filtering clouds overhead; the light is refracted as if through the bottom of a bottle made of colored glass. Darkness will soon seep in from the east, and what is decidedly the greatest wandering on our planet is about to begin. Every single night billions of tiny creatures, such as krill and various types of plankton, as well as millions of squid, rise up from the ocean deep to the nutrient-rich water of the surface. At dawn they drift back down to the dark.

Considering the time of year, Vestfjorden has been amiable and accommodating for half a day. But out here the weather has a short fuse. The wind is often strongest when the tide rises in the evening, as if it comes in with the water. In a matter of minutes Vestfjorden can fill with *poppel*, as some fishermen say when they want to describe sharp waves created by currents and winds moving in opposite directions.

We need to head back. But first Hugo tells me a story. In the 1970s, after he came home from Germany, Hugo played in a Tromsø band called Nytt Blod (New Blood). With its prog rock vibe and its over-the-top stage shows, the group was very popular. A big concert in Tromsø was supposed to open with the lead singer hanging naked from a cross.

"Not only that," says Hugo, "but the stage was supposed to be covered with smoke, and then the singer would slowly become visible as the smoke lifted." But the smoke machine short-circuited all the electrical equipment, and the singer ended up just hanging there in front of hundreds of audience members while the band was unable to play a single note. Finally he shouted: "Don't just stand there! Get me down from this bloody cross!"

And by the way, the band held their practice sessions in the Åsgård mental hospital.

Hugo nods and starts up the motor. After a moment he can tell that something is wrong. The outboard motor, which had been in the shop for repair, doesn't have the power it used to have.

The sound is more muted, and he notices a burning smell where he's sitting in the stern. The repair work clearly wasn't successful. We manage to get back to Skrova, but we'll have to take the motor to the shop again, and it isn't even on the island. This is more than a little annoying, since we have everything ready so we could spend several consecutive days fishing for a Greenland shark, and under extremely favorable conditions.

On the other hand, I'm not in a hurry. I haven't even bought a return plane ticket yet, and I've been stranded in far worse places than Skrova. Plus we probably have enough liver *graks* to lure Greenland sharks from all over Vestfjorden to wherever we like, if only we can get the outboard motor fixed.

33

For the next few days the weather is annoyingly stable, and it's aggravating to see the water so calm because we can't go out. Yet we're actually starting to get used to this kind of situation and quickly slip into the rhythm of Aasjord Station and the island.

An island is both real and its own metaphor, as the German author Judith Schalansky writes in her book *Atlas of Remote Islands*. I usually feel strangely free whenever I come to a small island like Skrova. It's as if life takes on a new rhythm, and my normal hectic pace feels distant and unimportant.

An island is a world in miniature and easy to master because the geography is clearly limited, as are the number of people and the stories to which you need to pay attention. Life seems simpler; a sense of perspective settles into your body. That's how Daniel Defoe describes the island life of Robinson Crusoe, who manages fine on his own as he personally moves through the various phases of civilization. He begins as a hunter and gatherer, then develops farming, raises animals, turns to architecture, slavery, war, and so on, making use of increasingly advanced technology. He eventually reaches the capitalist phase, with his balance sheet accounts and maximum-use view of the world.

On the island he also comes to understand who he really is, and he turns philosophical. Crusoe discovers that he can be happier alone on an island than anywhere else on earth. On the island he lacks for nothing. He's like a free-floating atom of noble gas, and he thinks of himself as emperor or king of his own realm. But he is cut off from humanity, and at one moment he sees his solitude as a punishment from God. He is thrown completely off balance when the parrot speaks to him. "Poor Robin Crusoe! Where are you? Where have you been?" But he is never truly afraid until he discovers footprints left by another person in the sand.

An island can be a paradise, but sometimes it's a prison. It's easy to have illusions about an island, imagining that everything will be great, that you're protected from the chaos and disruptions of the mainland. You may start to miss people and everything from which you've fled. A feeling of loneliness and isolation spreads over the whole island. You stop thinking of yourself as the emperor or king of a specific and limited realm. Instead, you feel trapped, surrounded by water on all sides. Maybe the autumn arrives, bringing darkness and silence. You long to leave nature behind and return to the city and other people. "But silence on an island is nothing. No one talks about it, no one remembers it or names it, no matter how strongly it affects them. It's the tiny peek at death that they're given while they're still alive."[16]

Some people turn their back on the world and put their trust in an even smaller island, a utopia where nothing can disturb them, an island that's small enough to encompass only their own personality and where they feel no longing for anyone else. An obsession may seize hold of some people; they change and start living a largely interior life. But either their personality is too small or the island is too big for any lasting happiness. That's the experience D. H. Lawrence describes in his mostly forgotten story "The Man Who Loved Islands."

The Atlantic Ocean is filled with mythological islands—places that have never actually existed except in the imagination of cartographers and poets. In the twelfth century, the famous Arabic geographer al-Idrisi reported there were twenty-seven thousand islands in the Atlantic. The truth is there are only a few dozen. So many expeditions have been sent out to discover islands that don't exist, yet were described in such detail by seafarers who claimed to have been there, even though no one was able to verify these fantasies of theirs. The descriptions were often so vivid that other seamen became convinced they'd been there too. Then they in turn made their own contribution by filling in the gaps about these islands of the mind.

At low tide I take walks on the island, usually along the shore's edge. Like most people, I have a personal relationship with the intertidal zone, having played there as a child. It's a pleasure to spend time in this space between sea and land. People taking walks like this seem compelled to collect small items to put in their pockets and then set them on the mantelpiece or kitchen windowsill. Smooth stones, shells, sculptural pieces of driftwood, or other things the sea has brought in. Maybe a message in a bottle will even turn up from the other side of the globe. During a certain period of my childhood I sent out bottled messages myself, saying that I was stranded on a desert island. And that wasn't really far from the truth, since I grew up in Finnmark.

Most Norwegians seek out the foreshore during vacation. They may have a cabin near the sea, or else they travel to a beach area in southern Europe. Nothing is considered more natural. Give a kid a plastic bucket and a shovel, and he can spend the whole day at the beach. He forgets about being cold or needing food. It's as if he belongs in that salt world of sand, waves, water, and rocks. Half naked, he plays in waves or builds dams, canals, and other edifices, totally focused, like a supervisory engineer. "History is a child building a sand-castle by the sea," the Greek philosopher Heraclitus (535–475 BC) is said to have written.

A bone, possibly from a moose or a reindeer, has washed ashore. All organic energy has been leached out of the pores through the microscopic tunnels in the bone tissue, which has now turned to mineral, hard and smooth. The pale-gray porous bone material weighs almost nothing, and it doesn't shine like it used to. The surface is dull and absorbs the light. All the gristle, flesh, and fat were merely a temporary cover that has now been cleansed away by the sea.

British scientists who have examined fossils from the Devonian period (ca. four hundred million years ago), when the first sea creatures crawled up on land, have made quite an amazing discovery. The jaws and teeth of the first land animals had developed to tear apart flesh, not to chew plants. The eyes sat on top of the head, and the animals lacked any sort of neck. So the first animals on earth were carnivores with fish heads who used their teeth to tear one another to shreds. The fish-heads ruled the land for eighty million years.[17]

You may find it difficult to get this image out of your mind once it takes hold.

On the seaward side of Skrova all of Vestfjorden stretches out before me. Straight ahead, to the southeast, I can see the islands of Steigen. A towering gray cloud cover creates a favorable backlight that is neither glaring nor blinding, but instead forms soft contours and muted contrasts. "Snotgreen, bluesilver, rust: coloured signs."[18]

If I climbed up the hill, I would see the island of Landegode outside Bodø, as well as Værøy to the southwest. Maybe even Røst, the island farthest out in Lofoten. There, a crew of Venetian sailors was stranded on their way from Crete to Flanders in 1431. Just after passing through the Strait of Gibraltar, the ship hit a rock. Repairs were done in Lisbon, and the ship continued its journey north. In the Bay of Biscay, it encountered a violent storm that broke the main mast and rudder, setting the ship adrift. In mid-December, the crew had to take to the leaky lifeboats and abandon ship. They fought the sea through snow and darkness for weeks, enduring hunger, thirst, disease, and exhaustion. On a single day, four crew members collapsed and died. From the ship they'd brought plenty of salted meat—but not enough wine.

The currents and winds carried the men farther and farther

north, into the wild nocturnal nothingness. They never expected to experience terra firma under their feet again. But then one of them caught sight of the archipelago that is Røst. The crew managed to find a beach, and there they landed on February 4, 1432. They were saved by the locals, and on return to Italy the captain, Pietro Querini, described them as "the most flawless people one can imagine." Their hospitality was quite without limits, and people from Røst are still said to be of a darker complexion than most others in Norway. At any rate, the Italians were given lots of dried cod (*stoccafisso*) to take home, and Italian chefs soon worked wonders with it. The export of fish from Røst to Italy has been going on ever since.

I decide not to climb up the slope to get a glimpse of Røst but instead continue along the beach. A few small pools have been left on shore by the tidewater, and in one of them a couple of young fish are swimming. A solitary seagull is sitting on a rock. When I pick up a clump of seaweed, swift sandhoppers scatter in all directions, even though there's no other place to hide.

The foreshore constitutes a border area between sea and land, but also between life and death. That was true, at least, in the Vikings' world because the intertidal zone was used as a place of execution. The methods varied. Many of those condemned to death were bound to a post, and the tides then did their job. Such is the case in Olav Tryggvason's saga, which recounts, with an exemplary economy of words, the fate of the worshippers of *seiðr,* a type of shamanistic sorcery, at Skratteskjær: "The king had them all taken out and brought to a skerry which was under water at high tide and he had them bound there. Eyvind and the others thus lost their lives. From then on, the skerry was called Scrat-Skerry," meaning wizard or troll skerry in Old Norse.[19]

When I was a kid, I once took part in something similar. We tied a friend to a post on the foreshore, and after a while everybody disappeared. I remembered I had to go home for dinner. A grown-up, who'd just happened to pass by, heard a boy shouting for help. The water had already reached up to his chest.

The intertidal zone isn't land and it isn't sea. It's something in between. All organisms that have adapted to these conditions have a foot in both worlds. One moment they're underwater, and the next they're on land that's nearly dry, maybe even under a scorching sun. They have to tolerate salt and water and rain, wind and dry spells. They have to protect themselves against everything that wants to eat them, from both the sea and the shore, and defend themselves from the birds overhead. Just like in the sea, here it's a matter of finding shelter and food, and also of hanging on tight whenever the waves come rolling in, waves that are often strong enough to move huge boulders.

For this reason, everything that lives on the foreshore must have extreme characteristics. Crabs, snails, and bivalves have almost impenetrable shells. Many species burrow into the sand when the tide comes in. The crab species *Hyas coarctatus,* commonly called the *pyntekrabbe* or "decorative crab" in Norwegian, covers itself with algae to make itself invisible. That doesn't sound especially "decorative." But on its shell are tiny hooks that can attach to kelp and sea grass, or to whatever drifts past, so the crab changes camouflage depending on its surroundings. Sometimes I think of this crab as a vagabond of the sea, at other times as a creature that merely wants to fit in.

Many types of snails live both on land and in the water, just like crabs. The hermit crab has no natural means of protection, so it carries around an empty seashell on its back. If danger arises, it scoots inside. The hermit crab is a squatter, constantly moving, because as it grows bigger, it has to change houses.

The common limpet slowly crawls around in search of food before clinging to rocks so tightly that tools are required to pry it loose. Limpets are edible, but I've never been served them anywhere in Norway. Scientists have discovered that the limpet's teeth, which are more than a hundred times thinner than a human hair, are made of the hardest biological material on earth. The fibers partly consist of a material called goethite, named for the German author Johann Wolfgang von Goethe.

The roe of sea urchins is also edible but present only during the winter, before spawning. At that time, it's possible to scrape loose the small eggs, which are the most powerful sea elixir. Crushed and empty sea urchins are often found scattered over the rocks. This is because, at low tide, crows and seagulls pick up the sea urchins and drop them onto rocks from a height of about fifty feet to smash them open and make the food inside easily accessible.

The sandhoppers are leaping among the rocks. At the edge of the low tide area, spawn hide in the kelp and seaweed and among the sea anemones. The sandhoppers hide between the tentacles of the dead man's fingers (*Alcyonium digitatum*), in the cilia of sea pens (*Virgularia mirabilis*), or even between the spines of sea urchins. The mouth of a sea urchin can open and close symmetrically, like ice-cube tongs, and is called in biology "Aristotle's lantern." It consists of eight identical parts connected in a circle, opening and closing like a little miracle of precision engineering. Hugo has long been planning to make a large-scale sculpture of the mouth of a sea urchin.

The wet white sand makes me think of something I once read about the early Christians, who were persecuted by the Romans and used secret signs to test whether they could trust one another. Whenever two people met, and either or both suspected they might not belong to the same religious sect, one of them would draw a long arc in the sand. If the other person drew a new arc, reversed and crossing the first one, the drawing would be complete. The finished image was a fish. Most of Jesus's first disciples were originally fishermen—before they became "fishers of men," as they called themselves.

The intertidal zone is wondrously wide. The moon and sun are practically lined up, so their gravitational forces are working in tandem. Ninety-seven percent of all the water on earth is found in the ocean, and all the water on earth is pulled in the same direction until stopped by land. The farther north in Norway you go, the greater the difference between high and low tide.

In the old days, coastal people would collect cockles and ocean clams on the foreshore. Ocean clams burrow into the sand but leave behind tiny holes. If you poke a stick into the hole, the clam will close up around it and you can pull it out. During the Lofoten fishing season several generations ago, both ocean clams and cockles were salted and used as bait.

A big jellyfish has recently washed ashore. The tentacles of a jellyfish trail behind, with thousands of tiny harpoons or barbs. It hunts by sinking down through the water column, with the tentacles spread out to the side so they can reach anything edible and stun it. When the jellyfish is alive, of course. It's impossible to say what killed this particular jellyfish, and I'm not planning to perform an autopsy. Jellyfish don't actually have a brain. Still, the whole creature reminds me of one big brain, carelessly ripped out of a human skull, trailing long threads made of nerves, arteries, and veins. Primordial brains floating in brine.

When philosophers want to challenge our perceptions they often ask: How we can be sure we are not a brain in a vat of liquid, being fed impressions of the world? Most often, their answer is: We can't.

My subconscious is also stretching out its tentacles to transport flotsam from the past up to the surface. One of my earliest memories is of the time I buried my hand deep inside a stinging jellyfish that had washed ashore on a desolate beach in eastern Finnmark. I probably thought it was actually some sort of jelly, or maybe that industrial waste product they packed in colorful boxes and sold to us kids at the time—a sort of sticky green or red slime that felt cold to the touch. I can still vividly recall the pain, which gradually increased, like nettles only much worse. In 1870, a stinging jellyfish that floated into Massachusetts Bay measured more than six feet in diameter and probably weighed more than a ton. In the southern Pacific Ocean there is a box jellyfish that can stop a grown man's heart in a matter of minutes.

The jellyfish in the Narcomedusae order is also said to be a tough little devil.

As a life-form, jellyfish have survived many mass extinctions. They can tolerate acidic water, they have few predators, and they drift around like zombies. Jellyfish need hardly any oxygen.

They have survived crises that have killed almost everything else on the planet.

Over the course of five hundred million years, five catastrophic mass extinctions have occurred on earth. The best known is the last one, called the Cretaceous-Paleogene (K-Pg) mass extinction event. It happened 65.5 million years ago. The reason it's so well known is that all the dinosaurs died, except for some small flying lizards.

An asteroid many times bigger than Skrova crashed into the Yucatán Peninsula at a speed of approximately forty-four thousand miles per hour. It's estimated that the explosion had the same force as hundreds of millions of hydrogen bombs. This was probably the worst day on earth since life began. Large parts of the American continent were pulverized, and the rest was left in a suffocating darkness of dust. Tsunamis occurred that were so powerful they changed the shape of the continents. Dust clouds covered the atmosphere so that the sun was not visible for months or years. Most of the forests that covered the earth burned down. Acid rain filled the sea. It was a sulfurous pool for several million years.

This wasn't even the worst mass extinction. The Permian-Triassic (P-Tr) extinction, which occurred 252.3 million years ago, was much more extensive. It may have been instigated by a huge volcanic eruption in the area that later became Siberia—because this was during the time when the supercontinent Pangaea was being formed.

The heat melted the permafrost. For millions of years, logs had been piling up in swamps and on the forest floor. The volcanic eruption started fires, and the earth became like a gigantic charcoal grill. New greenhouse gases got trapped in the atmosphere, and things snowballed, especially in the ocean, which began to release stored methane gas. The connections are uncertain, but regardless, the result was what scientists call the "Great Dying," or the "mother of all mass extinctions." Acidification and temperature increases in the sea led to the massive growth of poison-producing bacteria. About 96 percent of life in the ocean—which means the majority of all extant life—disappeared. The sea also lost its ability to bind carbon, emitting instead huge quantities of greenhouse gases.[20] The atmosphere was suffocated by smoke and gases. The sea was poisoned.

For several hundreds of millions of years, before any fish existed, trilobites dominated the ocean. There were many different varieties, and they ranged from a twenty-fifth of an inch to three feet in length. Some swam, others moved along the bottom. Some ate plankton, others larger prey. They looked like something between a crab and a lobster, but without legs or arms, although some were equipped with spears or sharp horns. Since they were so numerous and also protected by a shell, a tremendous number of these animals have been preserved in stone as fossils. In Norway alone, about three hundred fossil types of trilobites have been identified. But toward the end of the P-Tr extinction, this abundant old branch on the tree of life was abruptly cut off. Toward the end of the Great Dying, every single trilobite died, down to the very last robust individual. It took many millions of years for life on earth to get back on its feet.

Precursors to today's sharks swam around in the ocean 450 million years ago. About a hundred million years later, sharks had become so prevalent that scientists sometimes call this period the "age of sharks." Many species of sharks have also gone extinct, including the megalodon, a shark that was nearly sixty-five feet in length and weighed close to fifty-five tons. Its jaws were six feet across and filled with sharp teeth, each the size of a whisky bottle. Another interesting but much smaller creature that died out 320 million years ago is the *Stethacanthus,* also called the "anvil shark." On its back it had a helmetlike structure where the dorsal fin usually sits. This structure was filled with teeth pointing forward. Scientists can only speculate what these teeth were used for.

Sharks are the most hardy and adaptable of any large animal ever created by evolution. Some smaller species, like lampreys, horseshoe crabs, sponges, and jellyfish have been around for longer, but they seem somehow like anomalies or accidents. On the other hand, several types of very big sharks, like the anvil shark, goblin shark, frilled shark, and possibly even the Greenland shark, have been around for forever and a day. No other species can match this record. They have survived everything that has been thrown at them, including volcanic eruptions, ice ages, meteor impacts, parasites, bacteria, viruses, acidification, and other catastrophes that have led to mass extinctions. By the time the dinosaurs appeared sharks had already existed for eons. And they continued to thrive even as the dinosaurs and countless other species went extinct. There are still about five hundred different shark species swimming around in the world's oceans, and half of them have only been discovered in the past forty years. Some are rare and endangered. Others are very abundant and widespread.

Today eminent scientists at some of the world's foremost universities have been reporting in leading journals such as *Science* and *Nature* that we are in the early phase of the sixth mass extinction. The Great Dying took place over hundreds of thousands of years. Today species are disappearing at such a rapid rate that scientists compare it to the mass extinction that wiped out all the dinosaurs over the course of a few centuries. The driving forces behind this extinction of species are the loss of habitat, the introduction of nonnative species, climate change, and acidification of the ocean.[21]

We know what is causing this sixth mass extinction. We have been here only a few thousand years, but we've spread to all corners of the globe. We have been fruitful and multiplied. We have filled the earth and subdued it. We rule over the fish in the sea and the birds in the sky and every living creature that moves on the ground.

The chemistry of the sea is changing. Even in coastal areas that were previously teeming with life, there are now large anoxic dead zones. In the deep oceans these zones are even bigger. The sea is clearly not only our most important source of oxygen. It also binds enormous amounts of carbon dioxide and methane, the greenhouse gas that is twenty times more harmful.

The temperature and carbon content are rising in the atmosphere. The ocean's automatic reaction to this increase is to absorb more CO_2. In fact, the sea has absorbed half of all the carbon dioxide we have released since the Industrial Revolution began in the early 1800s.

When carbon dioxide is dissolved in water, the water becomes more acidic. The sea is approaching an acidity level that threatens bivalves, shellfish, coral reefs, krill, and plankton, on which the fish live. A more acidic ocean also affects fish eggs and larvae. Many species, such as kelp, succumb to rising temperatures, while others survive by moving north. But no one can escape the acidity. We probably won't experience this in our lifetime, but if the ocean becomes too acidic, most of the larger marine life-forms will die out. Negative trends can reinforce one another and cause an entire ecosystem to collapse. Life-giving plankton will disappear, while toxic plankton and jellyfish will survive, possibly along with the toughest sharks in the depths of the ocean.

When the balance is disrupted, various processes are set in motion. For example, as the sea becomes more acidic, its oxygen content also diminishes and its ability to bind new climate-impacting gases is reduced. The sea doesn't just keep absorbing carbon dioxide as the level increases in the atmosphere. Cold water can better hold carbon dioxide than warm water, just as a cold bottle of carbonated soda stays fizzy longer than a warm one. Eventually, as more carbon dioxide accumulates in the air, the ocean's ability to absorb greater amounts of it will decrease—and global warming will escalate. One of the worst scenarios a climate scientist can envision is when the sea begins to release the methane gas that is stored on the seafloor and in the ice. Then the snowball effect and feedback mechanism could run amok, and the warming could accelerate to a catastrophic degree.[22]

During all mass extinctions, including those initially caused by comets, the sea has played a key role. The major cycles and processes in the ocean take place so slowly that by the time problems arise, it's too late to do anything. The sea has a reaction time of about thirty years.

Acidification of the oceans has been going on since the nineteenth century, and in the best case scenario it will take many thousands of years before the sea returns to the same pH level it had at the start of the Industrial Revolution. Life in the sea as we know it will end. Possibly millions of life-forms will go extinct before we've even discovered them.

Plankton stores far more than half of the oxygen we breathe. If the plankton die, the earth is likely to become uninhabitable for humans. We will become, in the end, like the fish with the dulled eyes, gasping for air in the bottom of the boat. Obviously we could have taken better care of the ocean. But that's actually a rather self-centered statement, considering it's the sea that takes care of us. Climate changes, which the sea partially creates because of changes within its waters, will definitely have an effect on us. Over the course of several million years, life in the ocean may return and find a new, productive balance. We, on the other hand, can't press the pause button for a few million years. The relationship between us and the ocean is not like in some romantic love story, in which the mutual dependence is so strong that we can't live without each other.

Having said that, entire nations already relate to the sea in almost lovesick ways. This is something I discovered a few years ago when I arrived in La Paz, the capital city of Bolivia. In 1883, the Bolivians lost a war with Chile and had to relinquish their entire coastline. The fact that Chile took the ocean away from them has left a deep scar on the national soul. The Bolivians view it as the greatest injustice, and they still haven't given up reclaiming their coast via the international courts. While they're waiting for their coastal area to come back home, Bolivians try to keep up morale. They have a symbolic navy that bobs around in Lake Titicaca, and each year they celebrate Sea Day (Día del Mar) as a national holiday. Children and soldiers parade through the streets of the capital, for only what is lost is owned forever, though maybe not even that.

The sea will do just fine without us. We cannot survive without the sea.

34

On my way back from my walk along the shores of Skrova, I stop to have a chat with the ponies in the small glade where they're grazing. That's when my cell phone rings. It's my fiancée, who, by the way, has eyes that change color like the sea. Since my last visit to Skrova, we've found out that she's pregnant. She is seven weeks along, and everything is going well.

Both of us are happy and excited. We have started reading about fetal development, week by week. For a long time now I've also been reading books about fish, and the development of life on earth. I can't help connecting the two and making some observations.

Inside my fiancée a life is growing, surrounded by amniotic fluid. After seven weeks, an embryo bears a striking resemblance to a fish larva, and the similarity isn't merely superficial. The fetus has bulges or arches along its upper body. These are pharyngeal arches—gill arches in fish—which, over the course of the next weeks, will grow together to form the neck (larynx) and mouth. Right now the fetus has eyes on either side of its head, like a fish. The ears are way down on each side of the neck. What will become the nose and upper lip is now on top of the head. The indentation we all have in our upper lip is a result of this process. If something goes wrong and the merging isn't completed as it should be, the child may be born with a cleft lip or palate.

The embryo's organs and body parts move around almost like drifting continents, entering various stages of evolution. If it's a boy, what will become the testicles lie almost next to the heart. Gradually, as the embryo develops, the testicles slowly move down to where they belong. They need to be as cold as possible. In most fish, which are cold-blooded and have a constant temperature, this isn't important, so the gonads stay next to the heart.

Our ancestors came up on land, but we still have a lot of ocean inside us. The same muscles and nerves that enable us to swallow and speak were developed in the sea. Sharks and other fish use them to move their gills. Sharks and humans— Greenland sharks and us—have similar structures of neural pathways from the brain. Our kidneys and the interior of our ears are also souvenirs from our past in the ocean. Our arms and legs have developed from fish fins. Along with most animals and birds, we have quite a lot in common with fish.[23]

I don't tell my fiancée that we're going to have a fish, and of course we're not. But the creationists are right when they deny that we are descended from monkeys. Like the monkeys and all life on earth, we came from the sea. We are rebuilt fish.

35

Almost a week passes. Hugo and I are still unable to go out on the water, so I find myself simply hanging around. In my idleness I start to wonder what the hell we're really doing. And it's possible, since he has so much to do, that Hugo is starting to wonder what the hell I'm doing. We snap a bit at each other. Maybe we can't see much meaning in the whole project anymore. After all, he lives here and does what he does, while I keep coming to visit, though without considering myself to be a guest. Each time I return it feels as if I were gone for only a day. I feel like I slip right into place when I'm with Hugo and Mette, as if I've been silently adopted. But on some level I'm an intruder. I go in and out of their personal lives, bringing along my own habits, both good and bad. Even though Aasjord Station is considerably larger than many castles, the habitable part is no bigger than a small apartment. There's no such thing as an invisible guest. For good reasons the Arabs and many other people have a proverb that goes like this: After three days, guests, like fish, begin to stink.

There's no end in sight for Hugo and Mette in terms of the work they need to get done—carpentry and construction, permits and all sorts of arrangements—and I'm unable to offer them any real help. One day I hosed down the forecourt and wharf, when they weren't even dirty. And apparently I'll never learn to shut the stubborn door properly behind me, letting the heat escape along with Skrubbi the dog.

Hugo and I rarely argue, but it has happened before. Once it was over some minor issue, and I think both of us realized how trivial it was after the fact, but we had definitely insulted each other. Because of that "minor issue," we didn't speak to each other for two years.

Who ever said that minor issues aren't important? After wandering around the island these past few days, I end up feeling depressed and generally dissatisfied about several things in my life. I wish I had completed many more work assignments. Is it even possible to call what we're doing on Skrova "work"? And how many times am I going to fly to Bodø and disrupt Mette and Hugo's established routines?

One day I ask him bluntly: "Why do you really want to catch a Greenland shark?"

Hugo freezes and looks at me, his expression wary. "My father told me about lots of sea creatures when I was a boy, but it was the stories about the Greenland shark that stayed with me. The shark was so mysterious and creepy."

"But—"

"It's been at least thirty years since I started thinking about catching a Greenland shark by using the old methods. But now this project of ours has stripped the idea of all spontaneity. I'm doing it for my own sake—not so someone can read about it or so I'll be able to talk about it. For me, it's enough to see the shark. To experience the excitement when the shark comes up from the deep. And now that we've started this whole thing, there's no use stopping. We have to finish it. Sooner or later, that shark is going to come up."

In the Old Norse poem "The Lay of Hymir" (dedicated to the exploits of the giant Hymir and included in the *Poetic Edda*), there is a superhuman story about fishing. Hymir and Thor, the second most powerful god of Norse mythology, decide to go on a fishing expedition. For bait they use the head of an ox. The tale takes a dramatic turn when none other than the Midgard Serpent latches onto the hook. The Midgard Serpent, also known as Jörmundgandr, is not a fish but the sea serpent that grew so big it encircled the entire world (Midgard) and was able to bite its own tail. Thor and the serpent fight a fierce battle, but as thunder resounds in the sky, Thor hauls to the surface the ether-breathing Midgard Serpent. Thor gives a triumphant cry, but by then Hymir has had enough and he cuts the line before the serpent can be killed.

Thor later meets the Midgard Serpent again during the apocalypse called Ragnarok in Norse mythology. I won't spoil the story by telling you who won, but the event is known as the twilight of the gods, after all.

One afternoon Hugo and I drive the short distance to the western tip of Skrova, almost out to Elling Carlsen's old lighthouse. There we see some cormorants spreading their plumage, flapping their wings. Hugo says that's a clear sign it's going to rain tomorrow. I think that's just superstition, so I bet him a thousand kroner that it won't rain. He refuses to take the bet and seems a little annoyed. Maybe he suspects that I've checked the weather forecast, which of course I have. The next day, as predicted, there's not a drop of rain and hardly a cloud anywhere in all of Nordland.

Usually, when a question arises and we both think we know the answer, we behave very respectfully and one of us will ask: Can I go first? Okay, the other person then replies. But now we merely blurt out the answers, trying to beat the other person to the punch. Even when we start talking about food, conflict is in the air. Hugo accuses me of liking stew—as if that says it all—because it's what I've twice ordered from the store in Svolvær.

The tension rises in the middle of one of the Inspector Derrick TV shows that Hugo watches every afternoon. Maybe he wants to practice his German, or maybe he enjoys being mentally transported back to the same Germany he was living in during the 1970s. In the TV show, all the interiors and attitudes from those days are still intact. Hugo once ended up sitting right across from Horst Tappert, the actor who plays Derrick, at a dinner for artists on the island of Tranøy, just across Vestfjorden, where the German "friend of Norway" owned a house. Hugo found the actor to be a very engaging and courteous man. I point out that under no circumstances could the fictional Derrick be described in the same way. The inspector usually teeters on the border between moralizing common sense and scornful derision, which is also the case when it comes to his colleagues. He is ingratiating toward anyone from the upper classes, and he considers all Italians to be scoundrels from the second he meets them. Over the course of the 281 episodes in the TV series, Derrick has only two girlfriends. And after a short time, both women disappear without a trace. God knows what happened to them, but Derrick is on my list of suspects. The only reason I suggest that the oh-so-proper Inspector Derrick really is a deranged pervert is to provoke Hugo. A storm is slowly brewing. There may even be a fierce gale in the offing.

The next morning I'm sitting in Hugo's living room, writing an article on deadline. He's working on a painting in the next room. It's a commissioned piece depicting three famous islands—Ellefsnyken, Trenyken, and Hernyken—in the municipality of Røst. The islands are formidable mountain formations jutting up from the ocean. Hugo was supposed to have finished the painting several months ago, and in a few days an acquaintance is planning to take it to Røst. Someone who was born out there is going to hang the painting on his living-room wall. Hugo doesn't normally do naturalistic paintings, but a friend of his has commissioned this work, so he realizes that the mountains, at least, need to be recognizable.

Hugo is struggling because the mountains are so symmetrical and perfectly formed that it's almost too much of a good thing. Two of the mountains, which are side by side, are often compared to a woman's breasts. Next to them is a sharp peak. The sketches that Hugo has done look a bit forced. The light is often distorted out there, and it's difficult for an artist to depict the light as it reflects off the water and strikes the slopes. So Hugo keeps rubbing out what he has painted, trying to adjust the effect of the shadows and nuances. In the evening it looks great, but in daylight it has a glaring effect that robs the painting of all depth. When I first arrived, Hugo immediately asked me what I thought of the piece, and he seemed relieved when I confirmed his own opinion. I wasn't able to pinpoint any specific problems, but my impression was that the painting failed to meet his usual standard. In fact, as it looked now, it might almost be mistaken for the work of an amateur. And I've never thought that about any of his previous paintings.

"Exactly! That's precisely the problem," Hugo replied. He was not being sarcastic. He could see the problems with this piece better than I could.

The iconic and symmetric mountains rise up from the sea, side by side. Sometimes you simply have to accept that nature looks unnatural. The horizon is supposed to continue into infinity, which requires an illusion of unfathomable depth in the sky, and that can quickly lend the painting unintended religious overtones, and if this is exaggerated . . . I can see why Hugo is struggling.

But why does he have to turn up the radio so fucking loud? I turn down the volume every time he goes out for a short break. I can't write with that inane torrent of newscasts in the background, interrupted by dreary hit songs and ballad singers from the north. I have a deadline too. Or to be more accurate: I *had* a deadline. Now I'm working overtime on an article that has to go to print. Why can't Hugo put on that headset he's always wearing? He probably left it somewhere in the building.

Of course, this is his home, and I'm just a guest. But I'm also a friend. So when I'm trying to finish writing an article, isn't it okay for me to behave more like a friend than a guest, out of self-defense, so to speak? I notice that Hugo notices he's getting on my nerves. It's a red flag. Maybe even a convenient distraction from his own misery over those damn mountains, since he seems to be spending most of his time rubbing out what he'd already painted. Well, okay. Letting the meaningless chatter on the radio play in the background may be part of Hugo's creative process, at least at this stage of his work. Maybe the presence of a disruptive element allows him to push aside all other distractions, enabling him to work freely in some sort of jazzy way.

Every chance I get, meaning whenever he leaves the room, I turn down the volume almost as low as it will go. But Hugo always notices when he comes back, and he turns it up. We may be heading for a confrontation, which is the last thing I want, but the radio is driving me nuts. I can hardly write even a sentence or formulate a single clear thought.

The danger is that the situation could lead to a "discussion," which would constitute a total breakdown of the level of concentration I need to mobilize right now. So I try to make myself as uncommunicative as possible by sitting with my back turned, like a deaf shellfish. I don't answer when Hugo says something, and I hope that the back of my neck is radiating so much negative energy that he'll leave me alone. This tactic is risky, because it could easily provoke him, and that might serve to exacerbate the situation. We probably have twenty thousand square feet of space at our disposal in the building, which ought to be enough to keep out of each other's hair. But I need an Internet connection to check on some information before sending my text, and it only works here in the living room. The fact that the sun is shining on long, gleaming swells outside doesn't help our mood. We could have been out there fishing instead of sitting indoors, with each of us battling a deadline. If we had a boat, that is.

After I've turned down the radio for the third time, Hugo returns and starts to talk to me, addressing me in a way that forces me to reply. We're approaching what's called the *fallbrestet,* meaning the high-water mark of annoyance. If I don't watch out, he'll probably throw me out of the house. Then at least *he* can have some peace and quiet. Hugo asks me why I keep turning down the radio when he likes to have it on when he's working. And besides, who am I to talk about disruptions? Wasn't I the one who kept playing the same song over and over in the gallery last summer while he was hanging up paintings? For him, that process demands silence and an intense amount of concentration.

This is news to me. Apparently he gave me hint after hint about how nice it would be to work in silence, but I just kept playing my music. And several times I put on the same tune. Now he tells me that he's still allergic to the guitar riffs that start off that particular song. Why didn't he simply ask me to turn it off, the same way I'm begging him now? He claims he did.

I keep my mouth shut and retreat into my shellfish pose. I can tell he's close to losing his temper, but his sense of decency prevents him from throwing me out.

After several hours we both manage to meet our deadlines, and without allowing the situation to escalate into melodrama. By altering the previous image and creating softer transitions, and by changing the direction of the sunlight, Hugo successfully finishes his painting.

That same evening we start discussing something I've written, something he says lacks precision. What exactly he's referring to isn't important, but it has to do with northern Norway. I respond by demanding he display greater precision in his art, especially in his abstract paintings. And how precise are our actions out on the water? For instance, the triangulations we use. They're actually so imprecise that a little fog thirty miles away can make our points of orientation invisible. And besides, the strength of the current in Vestfjorden has fooled us many times. The line and bait disappear, even though we think they're lying nice and easy on the bottom. Yet before we know it, they're on their way north to Bear Island.

"What is precision in the art of painting?" I ask.

"Precision in the art of painting?!" Hugo exclaims.

I know that precision is not a central concept in his work.

"So maybe it has to do with the *opposite* of precision?" I go on.

"No, that's not it at all. It's not about precision, so it's not about the *opposite* of precision, either. It has to do with something totally different."

In the ensuing discussion I say that he worries too much about what we're going to do when we actually get a Greenland shark on the hook. We should be thinking more about whether we'll ever catch one at all. The way Hugo talks, it's more like a practical task that has to be carried out. Yet we both know there's more to it than that. The motive for this hunt of ours has a dark side. It may be taking place on a smooth surface that reflects the clouds, but underneath are hidden skerries and rocks, and visibility is limited. Clay and sediment are being whirled up from the bottom by what we describe as a monster.

In the real light, meaning in daylight out on the fjord, our "mission" shines with meaning. Yet it has clearly become an obsession, and we've invested a lot of ego into the project. We can't give up until we're looking into the whites of a Greenland shark's eyes, which of course will be obscured by a couple of long, dangling parasites.

What kind of idiotic, murderous mission have we embarked on? Is it a matter of satisfying our own curiosity? Of confronting our own fear? Of a hunting instinct that requires us to bring down the biggest prey we theoretically can manage, a sort of big-game hunt at sea? Is the myth of the monster slumbering in the deep an innate part of us, genetically inherited from the days when humans were prey for now-extinct predators, back when saber-toothed tigers dragged us half dead into caves to devour us in the dark? The battle between us and the crocodiles, hauling us down to their underwater lairs and tearing us to shreds? When I think about it, the rotating technique of the Greenland shark is actually reminiscent of the crocodile's.

We won the contest by acquiring a couple of extra pounds of cerebral matter, a jellylike gray substance that is on the brink of understanding almost everything, including how our own consciousness works. Yet the inheritance from our past is still present as a sort of deep memory. Why are the nature shows that Hugo watches on TV so full of beasts, with an ominous American voice-over that tries to fool us into believing someone is about to be swallowed by a horrifying monster?

Wasps are far more dangerous to humans than sharks. Globally sharks kill a total of ten to twenty people a year. During the same time frame, we kill about seventy-three million sharks. In spite of this, we consider the *shark* to be the dangerous predator. Hugo and I are not blind to the irony.

Every time a shark attacks a person, the news spreads around the world. People picture a cold-blooded murderer with lifeless eyes striking suddenly and silently, killing for the sheer pleasure of it. A jaw with several rows of a razor-sharp teeth shoots up through the water column to seize hold of the arm, leg, or waist of an unsuspecting swimmer. Fresh blood colors the sea red, and after a brief, unequal battle, the shark swims into the deep as it gulps down a body part or two. We fear the fact that they don't fear us.

Sharks will never win a popularity contest. Pandas, cats, puppies, dolphins, and baby chimpanzees are at one end of the spectrum. Sharks are at the very end of the other. Today, whenever a shark attacks someone, the incident is like an echo of a distant, primordial time when we didn't yet dominate the world with our superior technology. In a matter of seconds, our control over the world is wiped out. Suddenly we are not the one who kills but the one getting killed. The likelihood that this would happen to someone is almost nonexistent. But we fear landing down there in the cold deep, surrounded by creatures that will devour every last scrap of us, until everything about us totally disappears.

We are all going to disappear eventually. But on the dark bottom of the sea, where the fish and those crawling little animals are waiting, we would vanish so completely that the very thought is hard to bear.

Beginning in antiquity, explorers, geographers, and naturalists have gradually mapped the whole world. According to Dante, Odysseus did not go home to Penelope, as Homer claims. Odysseus wanted to press on, so he passed the Pillars of Hercules, left the Mediterranean, and headed westward on the open seas. According to Greek mythology, these pillars were raised to mark the border of the known, inhabited world. Even Hercules didn't dare travel beyond that point. But driven by curiosity, a thirst for knowledge, and an adventurous spirit, Odysseus set off into the unknown, as Dante writes in the *Divine Comedy* (ca. 1320). Dante sternly punishes Odysseus for this transgression, sending him almost all the way down to the bottom of hell, to the eighth circle, where he is permanently engulfed in fire.[24]

Only a few centuries ago, many still believed that there were people with dog heads, or with their face attached to their chest—or creatures that were a combination of scorpion, lion, and human being. Anyone who traveled far enough away from what was familiar and home ground risked encountering horses with wings, dragons breathing flames, and creatures with eyes that could literally kill. The existence of unicorns was commonly accepted. The oceans teemed with huge creatures possessed of the strangest traits and intentions.

The façades of medieval cathedrals swarm with both fantastical animals and demons, all of which were considered to be real. We have always feared the bestial, especially predators that could kill and eat us. Through our activities, we are eradicating other species at a dizzying speed because we have obtained hegemony on earth and mastery of the seas. We've come so far that there's almost never any question of a fair fight between humans and animals. These days, the real fight always takes place between people.

Today the wild animals are threatened. For the most part we encounter them only in zoos or on safaris, where people pay huge sums of money to catch a glimpse of big game on the savannas, maybe even through a telescopic sight. Seeing whales or sharks up close gives many people joy, but it also gives them status.

And by the way, sometimes whalers and whale watchers have come closer to each other than you'd think was possible. A few years ago, a boat packed with people from around the world was on a whale safari off the Norwegian island of Andøya. They were having a great time because there were lots of minke whales in the area. Their joy abruptly ended when a whaling ship appeared nearby. Before the very eyes of eighty whale-loving tourists, the ship's crew harpooned a minke whale. On their way back to shore, the tourists witnessed another whaling ship hoisting a minke whale on board, as blood ran in torrents. Those tourists, and especially the children, would remember that scene for the rest of their lives. In an interview in *Andøyposten,* the head of Norway's Småkvalfangerlag, an organization founded in 1938 to protect the interests of whalers, said, "It's important to point out that those who go out to watch whales are extreme opponents of whaling."[25]

Take note of one thing: These days most monster movies are no longer about wild animals. They're more often about perverted versions of ourselves, like zombies and vampires. Other threats to us usually come from outer space or, once in a while, from the sea. That's where something unknown still exists, something we can't completely control.

So what about Hugo and me? When putting on Brian Eno doesn't help, things aren't good. How about Robert Wyatt? Forget it. Not even with Robert Fripp on guitar. Right now even early Roxy Music comes up short.

Every year humpback whales change the long and complex songs they sing. The new tune is delivered over great distances from one group to another. Hugo and I don't update our music that often. The music we play tends to be forty years old. I try *Ummagumma,* Pink Floyd's double album from 1969. It's described as some of the strangest music they've ever released, and most of the band members have disavowed the album. But Hugo belongs to a small group of people who regard it as a brilliant masterpiece.

We have fried *klippfisk* for dinner. The *skrei* we caught some two months ago has now become the most perfect dried-and-salted cod. Hugo treated the *klippfisk* the way they did in the old days. He carried the drying fish inside and then took it back out so it wouldn't get too much sun or any rain. He took sheets of the salted and dried fish out to Vestfjorden to rinse them in pure seawater.

As the evening progresses, the general mood gradually improves, more or less keeping pace with the rising of the sea toward high tide. But when the waters turn and the tide goes out, the mood again sinks.

Before we go to bed, Hugo and I agree on one thing. Neither of us will mention the name "Greenland shark" until we've brought it up from the water. As if the mere words might evoke a curse. But don't think we're starting to have religious or superstitious notions about the shark. That's not what it is.

Shark worship does exist in other parts of the world. In Hawaii, the *'aumakua* was regarded as the most powerful guardian angel, and it often took the form of a shark. The Japanese thought of the shark as master of the ocean storms. In some island societies around New Guinea, a shark caller has higher status than all other members of the group. In the old days in Fiji, the islanders worshipped a shark god called Dakuwaqa as a direct ancestor of their highest chiefs. On the island of Beqa, the people still regard their shark god with such respect that just like Hugo and me, they never utter the name. For them it's all right to put it in writing.[26]

It's almost noon by the time I get up the next day. By then Hugo has been doing carpentry work for several hours. He comes into the building as I'm making myself a sandwich in the kitchen.

He asks me about something I thought I'd explained in great detail yesterday, because it was important.

"How scatterbrained can you be?" I say, and instantly regret my words.

At first Hugo doesn't reply, but two minutes later he asks me, with his head slightly bowed, what exactly it was I'd said when he came in. I deny saying what I'd said and at the same time apologize for saying it. Something is going on between us. It reminds me of the dregs left in the bottom of returnable bottles.

Fish have a so-called lateral line, a system of sense organs that keeps them from touching one another, even when they're swimming in large schools. Humans don't have that, so clearly it's time to take a break. I had planned on achieving close contact with the sea, but not by having Hugo heave me off the wharf.

As I'm walking through the station, I stop to take a closer look at the old diving gear the Finns left behind, hanging on the wall. The wetsuit is much too small, and a lot of the gear is missing, so you couldn't just jump into the water. Even though Hugo and I are getting on each other's nerves, I think: Why not keep things in the family? Hugo's daughter, Anniken, likes to go diving. She lives in Kabelvåg and could lend me the equipment I need. She might even come with me for a dive. It's been years since I've done any diving, and most of my dives were in distant places like Sumatra and Surabaya. To go diving in Vestfjorden suddenly seems like the only right thing to do.

But there's something else I need to do first.

36

I keep an old car in Skrova. I bought it last year because I have a house far out in the Vesterålen archipelago. Over the winter, water has leaked inside the car through the roof. The seats are damp, and there's water on the floor. An acrid, moldy smell permeates the whole vehicle.

I drive on board the ferry to Svolvær and continue along the sparkling fjords to Fiskebøl, where I catch another ferry to Melbu and Vesterålen. My route takes me across bridges and sounds, through Sortland and onward over a small mountain pass with hundreds of whispering streams, out to the seaward side in the municipality of Bø.

Here the landscape changes dramatically. The almost alpine, wide-open, and classic fjord landscape of northern Norway now looks more like the terrain in places like the Shetland Islands or Greenland. The green marine landscape is treeless, stripped bare in all its glory, with black peaks reaching many hundreds of feet into the air. The low flora hugs the ground, its colors blue, rust red, or pale green. Out here, the land has been ice-free for eighteen thousand years, longer than anywhere else in Norway.

Where the road ends, in Hovden, farthest out near the sea, my house stands on a green moraine ridge above a white sandy beach. The house is white, but every corroding nail is visible because the walls have been sprayed with salt from the ocean. I go into the living room and notice that the wallpaper on the ceiling is bulging. All it takes is the light touch of a fingertip to poke a hole through the paper. A steady stream of water spurts right into my face. I run to get a big pot, but it fills quickly, so I get a basin.

My great-grandfather built this house. Recently four other people and I purchased it along with the land, which covers more than twelve and a half acres. But property that is situated on the sea legally extends all the way out to where the steep underwater shelf begins. And from the water's edge on the shore below the house, there are well over three hundred feet to the drop-off. In other words, we own part of the sea. That doesn't feel quite right, but that's how it is.

The house is in a state of decay. A lot of water has collected along the pipes, and that's what has settled in the living-room ceiling. Water has also been blown in from the side. One of the winter storms has torn off a board from the façade. Water drips into the basin with sharp little plunks, breaking the muted, rolling rhythm of the waves washing ashore. *SWOOSH. Plunk. SWOOSH. Plunk.*

Water is in the process of taking over the house, which has withstood the sea and stormy winds for a hundred years. A little bailing was all my car needed, but the house should have been pumped out and then properly moored. The wind is blowing in from the sea, prompting a plaintive whistling sound from the roof and corners. Next to the house is an old well. It's full, and the water I drink from it tastes salty.

I get in the car to drive back to Lofoten. Condensation forms on the inside of the windows, but I hardly notice any difference when I wipe the water away, because there's sea mist on the outside. Through this blurry waterworld, I occasionally glimpse cormorants nesting on the cliffs as the waves crash against caves and channels. My car thinks it's a boat. I don't need any road signs, but I keep an open eye for the beam of lighthouses along the way. My body feels saturated with water, and I have a runny nose.

I phone Anniken, Hugo's daughter. She agrees to go diving with me.

37

Two days later Anniken and I tip backward out of a boat near Kabelvåg. Finally I'm underwater in Vestfjorden. I put my head down, lift up my feet, and let the weight belt do its job. Like a sea mammal, I glide down toward the bottom twenty-five feet below. I see an opening between two dense forests of large brown kelp and make my way through. The kelp, as tall as trees, has wide, flat, glossy blades that gently sway back and forth with the current up through the water column. The blades slide along my body without catching hold.

On the bottom I lie down to rest and look up. Above me I see ripples and a trembling blue light on the surface, which has now become the border to an entirely different world. On land, we have the sky above us and the sea below. Down here, I'm looking at a membrane so thin that it can't be said to have any tangible substance at all; it merely represents a direct transition to another element.

Most organisms on earth live down here. Very few species can live both on land and in the sea, and then only for brief moments. In theory, penguins are equally at home in both places, yet they're quite helpless on land. The same is true of seals, walruses, and turtles. Only amphibians and some snakes are masters of both elements.

At first the earth was covered by a shallow, lifeless ocean boiling with sulfur. Living cells arose and clumped together to form increasingly advanced organisms. Everything happened slowly until the point when life accelerated and shot out buds in all directions. For many billions of years, all life on earth existed in the sea. Creatures that are now extinct swam around weightlessly, breathing through gills and similar contrivances. It was only about 370 million years ago that the first living creatures hesitantly crawled up into shallow waters. They developed legs for walking and lungs for breathing. In the beginning they lived both in the water and on shore. Then they took their first steps to emerge fully and began to colonize the land. Some changed their mind and went back into the ocean.

Here the water is clean and clear because currents are constantly moving through the area. When storms come in, this strip of coast gets blasted right in the face. For most people, the sea seems foreign and threatening but also strangely intimate and familiar. If one blows air at a healthy baby's face, it will take a quick breath and close its mouth. This is one part of the phenomena called the Mammal Diving Reflex, or bradycardic response. Most healthy babies will hold their breath under water. Their heart rate will also slow down, their blood vessels will contract, and less oxygen will be transported to the extremities. The diving reflex decreases after six months, but it could be argued that babies are born to dive. All I hear right now is the sound of my own breath, a gaseous hissing when I inhale, deeper when I exhale, as the air combines with water and creates a wet gurgling sound. Breathing underwater reminds me of the wet, burbling sound of a baby's heart inside the womb, the way you hear it via ultrasound sensors. In the womb, we all were enveloped by salt water. Even our lungs were filled with salt water, until the last period before birth. We had no idea that anything else was possible until we were forced out into the dry world, surrounded by light, and a slap made us empty our lungs for the first time. And we screamed. No longer underwater, from now on we would have atmospheric oxygen as our life force. As if over the course of nine months we had mirrored and reexperienced the entire process that the creatures of the sea had gone through on their way from water to land. In the classic movie *The Abyss* (1989), in which an alien, otherworldly civilization finally emerges from the ocean's depths, the divers go so deep that they have to breathe a liquid blend of oxygen. "Your body will remember."

After lying on my back on the seafloor for a while, I continue swimming, moving away from the small clearing in the middle of the kelp forest. Finally I'm seeing the world through the optics of the ocean. A brown crab slips sideways toward a crack, then takes up position with its back against the wall, its claws raised. I pick it up, then set it down again and move on. A small school of what I think are Raitt's sand eels burrows into the sand. Starfish feel their way along the bottom of a knoll. The small fish keep to the kelp forest, along with all the camouflaged creatures that always hide there.

The water feels silky smooth against my body, even through the black rubber wetsuit. I keep swimming with the gentle current, among the silent, swaying kelp trees. I am weightless like water in the water. Not indistinct or a nothingness but like a drop *in* the ocean.

The sea anemones wave, the plumose anemones are passively decorative. A lumpfish scowls at me, its spines out. It has a silly pout and looks excessively arrogant. Then a small shoal of fry appears, sleek and silvery. With abrupt lurches the fish all dart in the same direction, though without a specific leader.

I'm actually in fairly shallow water, but I can still feel the pressure in my ears and sinuses. Jellyfish and many other species that live in the deep will explode if they're taken up to the surface—just as we would be crushed into a shapeless mass down below. Only thirty feet down in the water the pressure is double what it is on the surface. At a depth of sixteen hundred feet, the pressure is the same as fifty-one atmospheres. That's a heavy burden to carry. Divers who go down to great depths risk developing one or more nervous-system syndromes. They can become drowsy, they can fall asleep for short moments, or they may suffer from shaking, nausea, hallucinations, delusions, diarrhea, vomiting, and other symptoms that would be bad enough on the surface but are life threatening down in the deep. The pressure is so great that it's much harder for the lungs to move the oxygen mixture in and out. All deep-sea divers have to go through decompression, which can take days. Without it, the blood becomes fizzy, like champagne, which makes a person drunk in the absolute worst way. Clots that form in the blood, joints, lungs, and brain can lead to a painful death. That's how poorly adapted we are to the environment of the Greenland shark.

The queen of bubbles lives in an underwater cave. In the Sumerian *Epic of Gilgamesh,* which is the first great work of literature known to the world, the hero Gilgamesh seeks immortality and is told that it can be found in the form of a plant at the bottom of the sea. Gilgamesh ties stones to his legs and lets himself sink into the water. There he finds the plant that will give him back his youth. But he should have taken better care— because when Gilgamesh returns to the surface, a snake steals the sea plant while the hero is bathing.

Now is when it happens. A current along the bottom carries me off with a force you wouldn't think possible. There's no point trying to fight it. I'll just get whirled around. Instead, I keep my hands close to my body and let myself travel along, shooting through the water, farther and farther down, as I pass the most incredible sights. From now on, I'm swimming in the poem of the ocean—past sailing ships with torn sails to where the smiling sperm whale swims along the bottom in search of giant squids with eyes as big as plates and arms that sparkle; through the colorful violet forests of coral, where slimy eels slip in and out of skulls with clumps of seaweed on top. The current carries me along a deep sea trench to a big opening where the fin whale polyphonically sings a deep, plaintive ocean song. The low humming of the cod larvae can be heard from above, in between the trumpet fanfares of the seahorses, while the lobsters dance in circles around halibut and flounders that flap their tails to applaud. The wolffish, as usual, bear the faces of people I know. Ocean sunfish stand still in the water, lighting up the wide-open jaws of basking sharks. Stingrays fly past in formation like stealth bombers on a raid.

As I'm dragged farther down into the dark, I think that all hope is lost. How can I survive this pressure? My air should have run out long ago, but that doesn't happen. When I finally find myself in pitch-darkness, the strangest creatures start glittering. Between the shadows of those who have drowned, terrifying illuminations appear. The current takes me through the water for dozens of miles, always downward, until I hear a violent roaring, like inside a waterfall. I think that I must be approaching the huge sea gullet that is connected to the earth's core. The jet stream that is drawing me along the bottom seems to have carried me all the way to Moskstraumen, the maelstrom where the sea boils and churns more than anywhere else on earth. I am helplessly doomed.

There is a tornado of water. On the inside the wall is black, smooth, and shiny, with objects swirling around: the remains of shipwrecks, planks and logs, furniture, splintered crates, barrels, staves, and shattered old lifeboats. I grab hold of a barrel and cling to it, because it seems to be on its way up through the vortex.

I wake up on a rocky beach on the other side of Lofoten Point, near an abandoned fishing village. As I lie there, totally exhausted, I can still hear the clamor from Moskstraumen's rumbling mouth. Except for what I've just described, I remember nothing from that undersea journey through the ocean's navel.

38

When I get back to Aasjord Station after my underwater journey around Lofoten Point, Hugo and I quickly fall into the same negative rut. When he asks me whether I had a good dive, I nod affirmatively and tell him that Anniken says hello.

Late in the afternoon, Hugo finally gets the outboard motor back. We take the boat out on Vestfjorden to test it—and to toss a new bucket of liver *graks* into the water. By now the old *graks* has been diluted to a homeopathic concentration in Vestfjorden and the ocean areas beyond. The motor has been given a new oil pan and should be in top condition. Hugo gives it more throttle as we leave the bay, and his worried expression switches to relief.

We've passed the Skrova lighthouse and are level with Flæsa when we both notice something right in front of us. There is no mistaking what we're looking at. No other creatures can swim with such speed, and the oval white patches are very visible. We're in the middle of a group of killer whales, or orcas. Up ahead they are constantly breaching the surface with energetic leaps. Suddenly a calf appears right next to our boat. It sticks its head above the water, staring at us inquisitively with one eye. The calf is the same size as the boat, but two whales double its size are intently communicating with it. The whale's skin is like thick black vinyl, just like our RIB. Maybe at first glance the calf thought the boat was a sea animal and it wanted to become better acquainted. The adult whales call the calf back to the group, which is heading east, into Vestfjorden.

The orcas spurt up from the sea like plastic toys that have been held underwater in a bathtub. Then they dive back down as they continue moving forward at full speed, as if they have an appointment to keep and yet have enough time to play a little along the way. I've never seen any animal that is more impressive. Once, in the jungle in Africa, I had an experience that comes closest. A group of unruly chimpanzees came toward me through the overhead canopy, swinging from one tree to another, breaking branches and shrieking as they yelled rapid-fire messages to one another. Before anyone even had time to think, they were gone. The chimps sounded like a bunch of teenagers, racing past as they celebrated their high-school graduation. By comparison, the orcas are like Italian sports cars, but they're very much alive and everything about them gives the impression that they own the sea.

Five or six whales appear simultaneously and surround the boat. A few of them are very close. The group is heading in the direction of Tysfjord, and in the past, that was where the whales often went. Nine thousand years ago, during the Stone Age, people in Tysfjord made a rock carving that depicts a life-sized orca. For millennia the whales have been gorging on herring in Tysfjord every winter, but during the last few decades the herring haven't shown up for the appointment.

No two orcas have identical markings or dorsal fins; they are like fingerprints. The male has the larger dorsal fin, which sticks up nearly six feet from the body in a sharp triangle. The female's dorsal fin is narrower. The top resembles a wave, rather like in traditional Japanese paintings. Orcas are among the fastest swimmers in the sea. Only the speed of sailfish, swordfish, and possibly some smaller whales in the dolphin family can compare. But the orca is much bigger and stronger.

For fifteen minutes we follow the group, until the leader—most likely a female—suddenly signals to the others that they are done playing with us. All the whales dive at the same time and disappear. Hugo idles the motor, and we drift back in the same direction we came. We're now many sea miles northeast of the Skrova lighthouse.

—

Hugo hasn't seen orcas in Vestfjorden since 2002, and he's practically beaming with joy at their return. He once told me that if he had to choose what animal he'd like to be, it would have to be an orca. The eagle and the orca. Those are his animals. I remind him of this, then ask whether he wouldn't get tired of eating herring and mackerel. Hugo laughs and asks me what animal I'd like to be. I don't answer because it feels like the best ones are already taken.

We chat as we sit in the boat, which is riding on the choppy waves. Currents on the way out of and into Vestfjorden collide and have to figure out how to coexist, which can't happen without creating breakers and roiling waters.

Hugo tells me a story. No, actually it's more like he confesses to a shameful secret. In Steigen in the 1970s, young men fueled by testosterone would go out to shoot orcas with shotguns. They even bragged about it, Hugo says scornfully. This sounds undeniably primitive, but at that time the orca was blamed for the collapse of the herring stock. For all we know, some of the orcas in the group we saw may remember those incomprehensible encounters with people, because—like humans—they possess both intelligence and memory. The orca has the largest brain of all sea mammals except for the sperm whale, which, as we know, has the largest brain of all known creatures, both living and extinct. The orca's brain can weigh as much as fifteen pounds. It teaches its young to hunt, and every group is able to pass on particular customs from one generation to the next. Each clan has its own dialect, which is different in tone and frequency so they're able to recognize one another and separate their own from other, possibly hostile groups.

Orcas and humans have a life cycle that is very similar. The females, who often lead the group, become fertile when they're about fifteen years old. Until they reach their forties, they have at most five or six young. But they live until close to eighty.

"Do you know how the orca got its name?" Hugo asks. (In Norwegian it's called a *spekkhogger*, or "blubber-hacker.") "It can attack a blue whale, the world's biggest creature, which can weigh up to two hundred tons. Two of them grab the whale's flippers in their teeth. A third one bites into the soft part under the jaw. Then the rest of the group start to tear off the blubber from the blue whale," Hugo goes on. He adds that not even the great white shark has a chance against orcas.

Orcas hunt in groups, making use of cunning methods. They release big air bubbles under shoals of herring, or they take up a vertical position in the water and use their tails to create powerful, coordinated currents that render the herring disoriented and helpless. Orcas have also been filmed working together to make big waves that wash seals off ice floes.

In Steigen, Hugo has a pair of orca teeth. If you hold one in your hand, it's hard to let it go. It's as smooth as a conch shell and fills your clenched fist with its heft. Hugo tells me that when orcas set to work in a shoal of herring, thousands of herring heads are left floating in the sea. It's as if they've been sliced off with a razorblade, and it's not easy to comprehend how the orcas manage to do that.

An adult orca has hardly any natural enemies. But Hugo has read that it doesn't feel safe around pilot whales.

"The pilot whale will go after the young of both orcas and sperm whales. If a group of male pilot whales comes into the fjord, the orcas take off."

People in some parts of Nordland call the orca *staurkval,* or "stake whale," presumably because the huge dorsal fin looks rather like a stake. If so, it would have to be seen at great speed, and from the front. If you see the fin from a small boat, you should hang on tight, because orcas have been known to sink boats. Hugo tells me that a few years back, an orca started behaving very aggressively toward an eighteen-foot plastic boat right outside Skrova, almost at the same spot where we are right now.

What made it act like that? I wonder. Hugo is sure that stress and difficult circumstances can make an animal snap. For instance, he says, who would blame the orcas that live in the pens at Sea World in the United States for becoming aggressive and vengeful? These enormous predators, meant to roam freely in the open seas, get kidnapped and placed in a big pool. From then on they're trained to perform various stunts for a paying audience while dreadful pop music thunders between the tile-lined walls. As a reward for doing what the trainers/prison guards want, the whales get a bucket of herring. At night they're parked in small cubicles where they can hardly move, as if they were boats, while water is sprayed over their backs so they won't dry out. The tall dorsal fin no longer stands upright but begins to droop like a withered plant. The fact that intelligent creatures who are tortured in this way develop a desire to kill—which they have succeeded in doing a number of times—is not one of the mysteries of the universe.

In 2011, a group of activists tried to sue Sea World in San Diego, on the basis that whales have rights. The court dismissed the case. But in 2014, things went better for an orangutan in a zoo in Argentina. A court was asked to consider whether the orangutan named Sandra (age twenty-eight) was a thing or a person, which would then have consequences for how she could be treated. The fact that the orangutan was not defined as an animal had to do with the relevant interpretation of the law and the lawsuit. The orangutan was clearly not a *thing*. But not exactly a person, either. According to the Argentine newspaper *La Nación,* the court decided that she, or it, should be classified as a "nonhuman person." Even though she was not human, she possessed intelligence and an emotional life. The court concluded that if she was allowed to live in better conditions, she would clearly be happier. So the orangutan did have fundamental rights.

The experience with the orcas has undoubtedly raised our morale, and Vestfjorden seems once again an amazing place for adventures and fantasies. The sun has already set behind the Lofoten Wall. In the sky a lavender light has appeared, enameled green at the bottom. A salty new moon is rising between Skrova and Lillemolla.

Maybe this is what makes Hugo tell me about an experience he had the last time he was in Barcelona. His kids wanted to give him a surprise, so they paid for their father to go up in a hot-air balloon.

"We rose slowly over the city. It was early in the morning, but the city was awake, with all its sounds. At first we heard people talking, even music coming from the windows. When those sounds faded, we heard cars and traffic, the sound of machines, sirens, birds singing, all sorts of things. As we rose higher, more and more sounds got filtered out. Finally, when we got above the cloud cover, there was only one sound left. Do you know what the last sound was I heard as I looked down at the clouds over the city until it was totally quiet and the only thing left was the wind?"

I think for a couple of seconds, then shake my head.

"It was the sound of dogs," says Hugo. "Not barking or baying, but dogs communicating with each other over long distances."

We almost forget to throw the bucket of *graks* into the water outside Flæsa. It's still light enough to triangulate our position (using the Skrova lighthouse, the stone marker on Flæsa, and Steigberget, at the top of the Helldalsisen glacier). Even though it's snowing over Steigberget, we can still glimpse the mountain, so we know—with less than an impressive degree of precision— where we're going to start fishing tomorrow.

39

But where can wisdom be found?
Where does understanding dwell?
. . .
it cannot be found in the land of the living.
The deep says, "It is not in me";
the sea says, "It is not with me."[27]

Only a few long swells roll in from the sea. It's overcast, but the cloud cover is high and stable for as far as we can see to the west. The waves are long and heavy. Small round clouds shine like objects made of polished steel. Everything is looking good for a nice day at the Greenland shark banks near the Skrova lighthouse.

For bait we have some whale meat left over from the *skrei* party. We've left it out so it has spoiled. I fasten a big chunk of the meat on the hook and throw it overboard. The chain swiftly goes all the way to the bottom as Hugo's new Japanese reel sings. Because this time we're using a rod and reel, it should all be much easier now.

Hugo is wearing the special vest with suspenders. In the front, near his abdomen, the vest has a sort of shield made of thick plastic, with a hole in which to stick the fishing rod. This will make it possible for him to use his whole body, if necessary, to pull in the shark. The equipment is sticking several feet out from his groin, almost straight up in the air.

The fishing rod is also attached to the reel with strong metal clamps. If the rod goes overboard, the fisherman will go with it into the sea. This thought occurs to both of us, and it reminds Hugo of an incident in the 1980s. On a beautiful spring day, his family was out on the water on board a large fishing boat. Hugo got in a rowboat to go over to a small island to gather seagull eggs. There was only one place to go ashore, in a narrow bay. Because of the conditions on the seabed, the water always rushed in so that you were practically hurled ashore. The way back had to be carefully calculated, because he would have to surf the same violent undertow on the way out of the bay.

Hugo collected some gull eggs and got back in the rowboat, but he mis-timed the undertow as he headed back. The boat capsized. Just before Hugo hit the water, he heard his brother yelling: "There he *goes*!" The undertow pulled Hugo down into the sea, where he was tumbled around like a rag doll. He was drawn down to the bottom. Knowing a collision was imminent, he stretched out his arms in front of him, and the barnacles on the rocks shredded his hands. After impact he was immediately hurled back up through the water column like a projectile. Hugo managed to get hold of the rowboat. The bucket of eggs, none of which had broken, was bobbing in the water. When he got back to the fishing boat, everyone thought he was half dead because his face was covered with blood, but that was from his hands after he'd brushed his hair out of his eyes.

"Not only that," says Hugo. He is about to steer the story in another direction when he suddenly freezes. Something has taken a firm bite on the hook. And it can only be one thing. The RIB is being dragged backward against the strong current, and only a fish weighing many hundreds of pounds, or maybe even a ton, would be able to do that. Hugo is practically reclining, with his heels dug in against the pontoon to put up some resistance and not get dragged into the sea.

Could we at least catch some sort of shark? Is that too much to ask? It doesn't absolutely have to be a Greenland shark, I think. Recently an unknown type of shark was pulled up near Eggakanten outside Vesterålen. Even the scientists at Norway's Institute of Marine Research couldn't identify it. But we are confident this one has to be a Greenland shark. Hugo has direct contact. There's nothing between him and the Greenland shark, except for the line, with each of them at one end of it.

"Where's the knife?" Hugo asks me as the shark hauls the boat in the direction of Steigen. If the Greenland shark goes much faster, Hugo won't be able to stay in the boat and the knife would come in very handy. After a few minutes the speed decreases, and Hugo can quickly reel in some of the line by putting the setting in first gear. Intermittently, the shark yanks on the line for a minute or two, and the only thing to do is hang on. Once it makes several strong leaps, and when I move toward the back of the boat, where Hugo is, the bow starts to point alarmingly upward, so I have to retreat to keep the boat balanced. Then the Greenland shark settles down again, and Hugo can reel it in more. The shark is on its way up. If it wasn't properly hooked, it would have been gone by now.

Suddenly the shark makes a run for it, taking out a lot of line. I look at the reel, and there can only be a few dozen yards left when it stops. Hugo seems to be in full control, even though it must be hard for him to keep pulling in the shark. We don't converse, merely utter a few swear words to ourselves. There's nothing to say. We both know the disadvantages of using this type of method. If we were hauling in a line by hand, we could have tied the shark to a float and let it keep swimming on its own. We can't do that now that we're using a rod. The Greenland shark will pop up close to the boat, and the only thing we can do is . . . I look at Hugo and decide we'll just have to deal with whatever happens. If things really go wrong, we can always cut the line.

After half an hour the line straightens out. It won't be long now before the shark appears. Just below the surface the Greenland shark now starts to spin. The attached chain is only long enough to wrap around its body a couple of times, so the shark quickly reaches the line, which instantly breaks.

The sea is moving. I picture the massive gray back disappearing toward the bottom, with our hook in its jaw, plus twenty feet of chain dangling below. The life of this particular Greenland shark will never be the same after its encounter with us.

Everything is very still. In the background the beam from the Skrova lighthouse flashes. Several black-headed gulls have gathered near the boat. They can tell we have nothing for them, so they move off with the wind and the waves. Slowly and patiently the sea rolls onward, as it has always done before we were here, and as it will continue to do long after we're gone.

ACKNOWLEDGMENTS

My biggest thanks go to Mette Bolsøy and Hugo Aasjord. As all readers can see, this book could never have been written without our friendship. Thanks also to Anniken Aasjord. And thanks to others who have offered help, both big and small: Arnold Johansen, Leif Hovden, Frode Pilskog, Bjørnar Nicolaisen, Torgeir Schjerven, Inger Elisabeth Hansen, Sverre Knudsen, Anne Maria Eikeset, Håvard Rem, Inge Albriktsen, Hilde Linchausen Blom, Tora Hultgreen, Knut Halvorsen, and the team of Ronald and Kari Nystad Rusaanes (who often provided me with lodgings when I was in Bodø). To all those whom I haven't mentioned here but who deserve my thanks: Thank you so much!

My thanks to my Norwegian editor Cathrine Narum, whose enthusiasm, linguistic talent, and professional skills in general have greatly impressed me. That said, any factual errors are my own. Thank you to my fiancée, Cathrine Strøm. She has helped by giving me literary tips, by reading the manuscript, and by offering her support through the entire project. Finally, I send a greeting to our as yet unborn child, who was conceived between trips to the north and who will be born just about the time this book is published in Norway. May the sea treat you well.

NOTES

Summer

1. As a student I took a seminar about Rimbaud's poem, taught by the Norwegian poet Kjell Heggelund. When I quote from "Le bateau ivre," I've made use of the original French and a number of translated versions, without relying exclusively on any of them. The poem has been translated into Norwegian by Rolf Stenersen, Kristen Gundelach, Jan Erik Vold, and Haakon Dahlen (who has produced a Nynorsk version). These translations, in addition to several others, such as one by Samuel Beckett, are collected in *Å dikte for en annen. Moment til en poetikk for lesning av gjendiktninger. Berman, Meschonnic, Rimbaud* by Cathrine Strøm (thesis in comparative literature, University of Bergen, spring 2005). The English translation quoted here is by Wallace Fowlie (2005, as found online).

2. In 2003, the biologist E. O. Wilson initiated an Internet encyclopedia about life on earth, secretly hoping that all species would have been described within twenty-five years (www.eol.org). But Wilson had to admit that neither he nor anyone else had any idea how many species would be involved. Today science has identified only 1.9 million species, on land and in the sea, and most of them are tropical insects.

3. Much of the information about the shark's biology and social life is from Juliet Eilperin's book *Demon Fish: Travels Through the Hidden World of Sharks* (Pantheon Books, 2011), and *Sharks of the World* by Leonard Compagno, Marc Dando, and Sarah Fowler (Princeton Field Guides, Princeton University Press, 2005).

4. Or was there something hiding behind the ritual? Maybe the important thing was not to kill but to eat what you killed. In that case, the sacrifices can be interpreted as a celebration of the collective group. The ritual re-created the order and hierarchy of the universe.

It strengthened and confirmed the sense of community. The people not only shared food with one another, but also—through the sacrificial ritual—with the gods. The gods at the top, human beings in the middle, the animals at the bottom. Yet certain discoveries have been made that indicate cannibalism may have been involved in the culture that existed on Engeløya. These discoveries involved pots containing sawed-off bones. That immediately makes things much more complicated.

5. This information is from the BBC TV series *Blue Planet,* DVD no. 2, titled "The Deep," in which the TV crew follows a scientific study about the decomposition of a whale carcass.

6. Jonas Lie's story *"Svend Foyn og ishavsfarten,"* published in *Fortællinger og skildringer fra Norge* (1872). *Collected Works,* vol. 1 (Gyldendalske Boghandel, 1902), p. 148.

7. The quote is from Inge Albriktsen's article *"Da snurperen 'Seto' forliste—et lite hyggelig 45-års minne"* in *Årbok for Steigen,* 2006.

8. Later I found out enough about the life history of this boat to write a sort of obituary for it. It was built as a fishing boat by the Unterweser shipyard in Wesermünde in 1921 and christened the *Senator Stahmer* by the owner, Cuxhavener Hochseefischerei. In 1945, the boat was commandeered by the wartime German navy as an "outpost ship." It was sunk during a sabotage action in the harbor of Ålborg, Denmark. That same year, on Christmas Eve, it was raised from the sea. In 1947, the boat was put in service as the MS *Elsehoved* (Copenhagen), then sold to Govert Grindhaug in Åkrehamn, Norway, near Kopervik, in 1950, whereupon it was rechristened the *Seto.* In 1952, the boat went down near Gulleskjærene, forty-four nautical miles north of Bodø. In other words, it sank outside of Steigen. It was then that Johan Norman Aasjord, son of Norman Johan Aasjord (Hugo's great-grandfather), saw his chance and bought the boat as it lay underwater. He raised it out of the sea himself. Aasjord repaired and rebuilt the boat as a herring seiner. Off season, when the herring fishing was over, the *Seto* served as a freighter to the Continent, returning to Steigen with its holds filled with liquor. On February

26, the boat capsized and sank in deep water outside Runde, during the winter herring season. And that is where the boat still lies, since this third sinking proved to be its last. Http://www.skipet.no/skip/skipsforlis/1960/view?-searchterm=norske+skipsforlis+1960.

9. Not to be confused with his ancestor Gerhard Schøning (1722–1780) from Lofoten, who became a teacher at Trondheim Cathedral School, professor at Sorø Academy, and the director of National Archives and Records in Copenhagen. Some consider this Gerhard Schøning to be Norway's first historian because of the academic works he wrote.

10. The names reveal where the fishing village owners were from. Many came from southern Norway, but at this time it was also typical for Norwegians to be of Danish, German, or Scottish heritage—with names like Walnum, Dybfest, Zahl, Rasch, Dreyer, Blix, Lorentz, Falch, Bordevick, Dass, Kiil, and others. They regarded themselves as members of the European upper class and frequently went on shopping trips to the Continent, where they purchased everything from large quantities of Bordeaux wine to chandeliers, grand pianos, carpets, and draperies. Their birthright gave them certain privileges, and they were entitled to decide where the common people could fish, how much debt could be heaped upon others, and which servant girls they would bed. A minority of these men were compassionate patriarchs who, during times of crisis, held a protective hand over their subjects. The parish priest Petter Dass (1647–1707), author of the famous *Nordlands trompet* (The Trumpet of the North) and other works, was not one of them.

11. Christian Krohg, "*Reiseerindringer og folkelivsbilder,*" in *Kampen for tilværelsen* (Gyldendal, 1952), p. 306.

12. Claire Nouvian, *The Deep: The Extraordinary Creatures of the Abyss* (University of Chicago Press, 2007), p. 18. This is an amazing coffee-table book with hundreds of photographs of life-forms that live in the deep.

13. By the age of twenty-four, Michael Sars had already self-published a scientific treatise titled *Bidrag til søedyrenes naturhistorie* (Bergen, 1829).

14. Truls Gjefsen, *Peter Christen Asbjørnsen—Digter og folkesæl* (Andresen & Butenschøn, 2001), pp. 236–42.

15. *Norsk biografisk leksikon,* https://nbl.snl.no/Peter_Christian_Asbjørn sen.

16. Quoted from *Norsk biografisk leksikon,* https://nbl.snl.no/Michael _Sars.

17. Four years later, G. O. Sars published several of his father's discoveries, as well as his own, in the book *On Some Remarkable Forms of Animal Life, from the Great Deeps off the Norwegian Coast. Partly from the Posthumous Manuscripts of the Late Professor Dr. Michael Sars* (Brøgger & Christie, 1872).

18. Jonas Collin (ed.), *Skildringer af naturvidenskaberne for alle* (Forlagsbureauet i København, 1882).

19. Ibid., *"Havets Bund,"* P. H. Carpenter, p. IIII. The English was translated from the author's Norwegian.

20. Wendy Williams, *Kraken: The Curious, Exciting, and Slightly Disturbing Science of Squid* (Abrams, 2010), p. 83. Williams's excellent book on squid is my main source for facts on this species.

21. See Tony Koslow, *The Silent Deep* (University of Chicago Press, 2007).

22. Jonathan Gordon, *Sperm Whales* (World Life Library, 1998).

23. Philip Hoare, *The Whale* (Harper Collins, 2010), p. 67.

24. Torgeir Schjerven. This contribution to the description of wonderment is a verse from *Harrys lille tåre* (Gyldendal, 2015).

25. LynneRose Cannon, ed., *Herman Melville: The Dover Reader* (Mineola, NY: Dover Publications, 2016), p. 296.

26. Ibid.

27. The whole story of commercial whaling, including the role played by science, is described in detail in D. Graham Burnett's book *The Sounding of the Whale: Science and Cetaceans in the Twentieth Century* (University of Chicago Press, 2012). The Russian navy alone took twenty-five thousand humpback whales in two seasons, in 1959 and 1960. Even before whaling became mechanized, the whale populations were so heavily exploited that they almost disappeared for good. Beginning in the early seventeenth century, Dutch, British, Ger-

man, and Danish ships caught tens of thousands of bowhead whales (*Balaena mysticetus*) in Svalbard, until there were very few individuals left by 1670. This whaling endeavor was effectively described by Frederick Martens, a surgeon from Hamburg who served on a whaling ship in 1671. His book about his experiences became well known when an anonymous English translation was published in 1694, titled *A Voyage into Spitsbergen and Greenland.* The bowhead whale, which can weigh up to eighty tons, belongs to the right whale family (so named because they were the "right" whales to catch). To impress the female bowhead whales, the males sing polyphonically, and they never repeat the same song two seasons in a row.

Over the course of about sixty years, up until 1967, as many as 450,000 blue whales may have been caught in the Antarctic Ocean alone. The Russians did not report all their catches, so it's impossible to know how many blue whales they took. Whaling was also the basis for the creation of many Norwegian fortunes.

28. In 1920, the Danish doctor Aage Krarup Nielsen sailed on a whaling ship from Norway to Deception Bay. The voyage is described in *En hvalfangerfærd* (Gyldendal, 1921). Nielsen claimed that the smell of poisonous gases used by the Germans in World War I was like a "plaything" compared to the stench in Deception Bay.

29. *New Scientist,* December 10, 2004, http://www.newscientist.com/article/dn6764.

30. George Orwell, "Inside the Whale," in *Essays* (Penguin, 2000), p. 127.

31. Lars Hertervig. *Lysets vanvidd,* documentary film shown on Norwegian TV, 2013.

32. From the poem "Bølgje," by Halldis Moren Vesaas.

33. From Rimbaud's "Le bateau ivre."

34. Fridtjof Nansen, *Blant sel og bjørn* (Jacob Dybwads Forlag, 1924), pp. 238–39.

35. Levy Carlson, *Håkjerringa og håkjerringfisket, Fiskeridirektoratets skrifter,* vol. 4, no. 1 (John Griegs Boktrykkeri, 1958).

36. Erik Pontoppidan, *Det første forsøg paa Norges naturlige historie, forestillende dette kongerigets luft, grund, fjelde, vande, vækster, metaller,*

mineraler, steen-arter, dyr, fugle, fiske, og omsider indbyggernes naturel, samt sædvaner og levemaade. Oplyst med kobberstykker. Den vise og almægtige skaber til ære, såvel som hans fornuftige creature til videre eftertankes anledning (Copenhagen, 1753; facsimile edition, Copenhagen, 1977), vol. 2, p. 219.

Autumn

1. Perhaps appropriately enough, a certain confusion exists regarding Aeolus, or Aiolos, Greek god of the wind. He appears in three different genealogies. In one version, he is the son of Poseidon. In *The Odyssey* (book 10), Aeolus is said to be the "keeper of the winds" and the son of Hippotes. Aeolus gives Odysseus a sack filled with all the winds so he can sail home on a steady west wind. But Odysseus's men think the sack contains worldly treasures, so they open it and release a hurricane. They are blown back to the island of Aeolia, where Aeolus refuses to help them a second time.

2. *A Voyage to the North, Containing an Account of the Sea Coasts and Mines of Norway, the Danish, Swedish, and Muscovite Laplands, Borandia, Siberia, Samojedia, Zembla and Iceland; with Some Very Curious Remarks on the Norwegians, Laplanders, Russians, Poles, Circassians, Cossacks and Other Nations. Extracted from the Journal of a Gentleman Employed by the North-Sea Company at Copenhagen; and from the Memoir of a French Gentleman, Who, After Serving Many Years in the Armies of Russia, Was at Last Banished into Siberia* (first published ca. 1677). In *John Harris Collection of Voyages and Travel,* vol. 2 (Navigantium atque Itinerantium Bibliotheca, London, 1744).

3. This information is from Arne Lie Christensen's book *Det norske landskapet* (Pax, 2002), p. 75.

4. Not all water on earth comes from space. We know this because the chemistry of meteor water is slightly different from the rest of the water. The hydrogen is of a heavier isotope. Only half of our water could have come from comets and other objects that have crashed on earth. The rest was probably here from the very beginning, in

material that formed what would become our globe. In other words, a large part of the water on earth is more than 4.5 billion years old.

5. Robert Kunzig, *Mapping the Deep* (Sort of Books, 2000), ch. 1, "Space and the Ocean."

6. It is estimated that there are approximately five hundred billion galaxies in the universe, each having billions or many thousands of millions of stars. In 2013, astronomers at the University of Auckland, using new technology, increased the number of "earthlike" planets in the Milky Way. The old estimate was seventeen billion. The new estimate is more than five times that number (one hundred billion).

7. NASA, and the scientist teams with whom they cooperated on this project, analyzed the data from the Kepler space telescope over a period of four years. They were looking for a planet orbiting around a sun at a distance that would make the planet habitable. So far, the planet that most resembles the earth is in the constellation Cygnus, which is fourteen hundred light-years away from our solar system. It was christened Kepler-452b.

8. The history of Norwegian lighthouses during this period, including the role of the Mork family, is best described in Jostein Nerbøvik's book *Holmgang med havet, 1838–1914* (Volda Kommune, 1997).

9. *Ny illustreret tidende*, Kristiania, June 26, 1881, no. 26, pp. 1–2.

10. Christoph Ransmayr, *The Terrors of Ice and Darkness,* translated from the German by John E. Woods (Grove Press, 1991), pp. 113–14.

11. Gunnar Isachsen, "*Fra Ishavet,*" *Særtryk av det norske geografiske selskabs årbok 1916–1919,* p. 198.

12. Jostein Nerbøvik, *Holmgang med havet,* p. 312.

13. http://da2.uib.no/cgi-win/WebBok.exe?slag=lesbok&boktid+=ttlo4. Frode Pilskog of the Dalane lighthouse museum identified Wiig as the designer of the Skrova lighthouse. He also sent me copies of the original drawings that are signed "Wiig."

14. The quote is taken from Alexander Kielland's novel *Garman & Worse,* originally published in Norway in 1880.

15. Bjørn Tore Pedersen, *Lofotfisket* (Pax, 2013), p. 109.

16. The English is translated from the author's Norwegian, based on the Swedish translation (the work was originally written in Latin). *Historia om de nordiske folken* (Michaelisgillet & Gidlunds Förlag, 2010).

17. The source of this story is a Welsh cleric by the name of Giraldus Cambrensis (1146–1223). He supposedly saw little geeselike birds hatch from the fruits of trees in Ireland near the sea.

18. During the battle of Actium, echeneis fish, or sharksuckers, reportedly seized hold of the ship belonging to Mark Antony's admiral. That was why Gaius Octavius (who became the emperor Augustus) was able to attack him so quickly. On another occasion, the fish were said to have stopped a ship with four hundred rowers. In addition, eating a "ship holder" can prove fatal. Olaus Magnus, book 21, ch. 32.

19. Ibid., book 21, ch. 41.

20. Ibid., book 21, ch. 5, pp. 987–88.

21. Ibid., book 21, ch. 35.

22. The "big Norwegian serpent" or dragon, which Norwegian fishermen had described to Olaus Magnus, may have been inspired by the Midgard Serpent. According to Norse mythology, Odin threw the Midgard Serpent out of Åsgard, home of the Æsir gods. At the bottom of the sea, the serpent grew so big that it eventually encircled the entire earth—just as Oceanus did in early Greek mythology. Thor once caught the Midgard Serpent on a hook when he was out fishing. According to the *Elder Edda,* when Ragnarok occurs, Thor and the Midgard Serpent will fight in a battle of the giants, from which no one will emerge alive.

23. Erik Pontoppidan, *Det første forsøg paa Norges naturlige historie, forestillende dette kongerigets luft, grund, fjelde, vande, vækster, metaller, mineraler, steen-arter, dyr, fugle, fiske, og omsider indbyggernes naturel, samt sædvaner og levemaade. Oplyst med kobberstykker. Den vise og almægtige skaber til ære, såvel som hans fornuftige creature til videre eftertankes anledning* (Copenhagen, 1753: facsimile edition, Copenhagen, 1977), vol. 2, pp. 318–40.

24. Ibid., p. 343.

25. *Kongespeilet* (author unknown), from the mid-thirteenth century, is considered perhaps the most significant work from the Middle Ages in Norway. In the book, a father tells his son about everything that exists in the world. The father says that in the sea off Greenland, there are both mermaids and sea trolls called *havstramb* (mermen). "Whenever these trolls have been seen, people have also been convinced that a storm at sea will follow . . . If the troll turns toward a ship and dives, people have known for certain [that they are in for a storm]. But if the troll turns away from the ship and dives in another direction, then there is hope that the crew will not be harmed, even if they encounter huge waves and a violent storm." *De norske bokklubbene*, 2000, pp. 52–53.

26. Pontoppidan, *Det første forsøg*, vol. 2, p. 317.

27. Bjørn Tore Pedersen, *Lofotfisket*, pp. 109–10.

28. A. C. Oudemans, *The Great Sea-Serpent: An Historical and Critical Treatise* (Leiden and London, 1892).

29. Olaus Magnus, *Historia om de nordiske folken*, book 21, ch. 34.

30. If you give a squid the choice between five boxes marked with different symbols, and you hide a crab in one of the boxes, the squid will quickly learn which symbol signifies the crab. If you put the crab in a different box, the squid will realize that the symbol for the crab has changed. Wendy Williams, *Kraken*, pp. 154–58.

Winter

1. Scott Stinson, "Skipper Uses Knife to Kill 600-Kilo Shark," *National Post*, November 2, 2003.

2. Einar Berggrav, *Spenningens land* (Aschehoug, 1937), pp. 36–37.

3. Mark Kurlansky, *Cod: A Biography of the Fish That Changed the World* (Penguin, 1997), pp. 50–51.

4. Richard Ellis, *The Great Sperm Whale: A Natural History of the Ocean's Most Magnificent and Mysterious Creature* (University Press of Kansas, 2011), pp. 123–25.

5. Russian fishermen think that seismic shooting off northwest Russia

destroyed their cod fishing grounds in the 1970s and '80s. Norwegian coastal fishermen have tried to prevent the seismic ships from entering their fishing grounds, but the coast guard boarded the fishing boats and expelled them from the area. The oil industry has financed the seismic shooting, and also uses the military—meaning the coast guard—as a security service when the fishermen stage demonstrations. This in an area where the authorities have actually decided that for the time being no oil drilling activities should take place.

6. Frank A. Jenssen, *Torsk: Fisken som skapte Norge* (Kagge Forlag, 2012), pp. 52–53.

7. Philip Hoare, *The Whale*, p. 34.

8. There are two Skrova songs. One is titled "Skrova-sangen" (The Skrova song) and was written by Wilhelm "Ville" Pedersen around 1950. It should probably be considered the official Skrova song. The other goes by the name of "Se Skrova-fyret blinker" (Look how Skrova's lighthouse flashes) and was written in 1949 by Herleif Peder Risbøl. It made its first appearance in a revue at the Youth House.

9. Olaus Magnus, *Historia om de nordiske folken,* book 21, ch. 2, p. 984.

10. Johan Hjort, *Fiskeri og hvalfangst i det nordlige Norge* (John Griegs Forlag, 1902), p. 68.

11. Johan Hjort later collaborated with the foremost British whale researcher, John Murray, who had participated on the first major and legendary deep-sea expedition, on the HMS *Challenger.* The sailing ship left Portsmouth Harbor in 1872 and sailed the world's oceans for four years. During the course of the voyage, they discovered well over four thousand new species. In 1910, Hjort and Murray both joined the expedition on the steamship *Michael Sars.* The voyage took them from the North Atlantic to the coasts of Africa. Hjort and Murray discovered more than a hundred new deep-sea species. They also found out that fish and other creatures in the deep often produce light by using chemicals and bacteria (bioluminescence). They described their discoveries in the book *The Depths of the Ocean* (1912). A Norwegian translation was published at the same time: Sir John Murray and Dr. Johan Hjort, *Atlanterhavet. Fra overflaten til*

havdypets mørke. Efter undersøkelser med dampskipet "Michael Sars" (Aschehoug, 1912).

Spring

1. Marine biologist Dag L. Aksnes has researched the phenomenon and led the research project "Coastal Water Darkening Causes Eutrophication Symptoms." A popularized version of the findings was published in the Norwegian journal *Naturen:* Dag L. Aksnes, "*Mørkere kystvann?,*" no. 3, 2015, pp. 125–32.

2. Per Robert Flood, *Livet i dypets skjulte univers* (Skald Forlag, 2014), p. 59.

3. http://onlinelibrary.wiley.com/doi/10.1002/2014GL062782/ abstract?campaign=wlytk-41855.6211458333.

4. Sigri Skjegstad Lockert, *Havsvelget i nord. Moskstraumen gjennem årtusener* (Orkana Akademisk, 2011), p. 111.

5. Edgar Allen Poe, "A Descent into the Maelström," in *Poetry, Tales, & Selected Essays* (Library of America, 1996).

6. Jules Verne, *Twenty Thousand Leagues Under the Sea* (Project Gutenberg, 2002, ch. 22, https://www.gutenberg.org/files/2488/2488-h /2488-h.htm).

7. Christian Lydersen and Kit M. Kovacs, "*Haiforskning på Svalbard,*" in *Polarboken 2011–2012* (Norsk Polarklubb, 2012), pp. 5–14.

8. Werner Herzog, "Minnesota Declaration: Truth and Fact in Documentary Cinema," a speech delivered at the Walker Art Center, Minneapolis, Minnesota, April 30, 1999.

9. Donovan Hohn, *Moby-Duck: The True Story of 28,800 Bath Toys Lost at Sea* (Viking, 2011).

10. *The Guardian,* March 8, 2013.

11. Recently, the Norwegian Department of Fisheries issued a controversial permit for the coast of Nord-Trøndelag. Fisheries journal *Fiskaren,* June 17, 2015, p. 5.

12. Gustav Peter Blom, *Bemærkninger paa en reise i nordlandene og igjennem Lapland til Stockholm i aaret 1827* (R. Hviids Forlag, 1832), pp. 77–78.

13. Svein Skotheim, *Keiser Wilhelm i Norge* (Spartacus, 2001), p. 168.

14. I have taken much of the information about the age of the earth and efforts to establish a precise age—from Bishop Ussher's days to modern times—from Martin J. S. Rudwick's eminent book *Earth's Deep History: How It Was Discovered and Why It Matters* (University of Chicago Press, 2014).

15. Ivar B. Ramberg, Inge Bryhni, Arvid Nøttvedt, and Kristin Rangnes, eds., *Landet blir til. Norges geologi,* 2nd ed. (Norsk Geologiske Forening, 2013), pp. 89–90.

16. Roy Jacobsen, *De usynlige* (Cappelen Damm, 2013), p. 97.

17. http://www.lincoln.ac.uk/news/2013/05/691.asp.

18. James Joyce, *Ulysses,* ed. Jeri Johnson (Oxford University Press, 1998), p. 37.

19. "Olav Trggvasons Saga," in *Heimskringla; or, The Lives of the Norse Kings,* by Snorre Sturlason, edited with notes by Erling Monsen, and translated into English with the assistance of A. H. Smith (Dover, 1990), p. 167.

20. Elizabeth Kolbert, *The Sixth Extinction: An Unnatural History* (Henry Holt, 2014).

21. The latest research report to be published on the subject was in the journal *Science Advances* from June 19, 2015, titled "Accelerated modern human-induced species losses: Entering the sixth mass extinction."

22. Tim Flannery, *The Weather Makers: How Man Is Changing the Climate and What It Means for Life on Earth* (New Atlantic Press, 2005). As the sea warms up, its ability to spread warmth down through the water column is also disrupted. The temperature difference between the three main layers of water increases, while the exchange between them diminishes. The warm water does not make its way into the deep, which further increases the warming of the surface. Fifty-five million years ago, the entire ocean was so warm that almost all life in the deep, which can survive only in cold water—and the Greenland shark is a good example—died out.

23. Neil Shubin, *Your Inner Fish: A Journey into the 3.5-Billion-Year History of the Human Body* (Pantheon Books, 2008).
24. Dante Alighieri, *The Divine Comedy,* song 26.
25. *Andøyposten,* July 3, 2006.
26. Juliet Eilperin, *Demon Fish.*
27. The Book of Job, 28:12–14, New International Version.

A NOTE ABOUT THE AUTHOR

Morten Strøksnes is a Norwegian historian, journalist, photographer, and writer. He has written reportage, essays, portraits, and columns and reviews for most major Norwegian newspapers and magazines. He has published four critically acclaimed books of literary reportage and contributed to several others.

A NOTE ON THE TYPE

This book was set in Adobe Garamond. Designed for the Adobe Corporation by Robert Slimbach, the fonts are based on types first cut by Claude Garamond (ca. 1480–1561).

Composed by North Market Street Graphics,
Lancaster, Pennsylvania

Printed and bound by LSC Communications,
Harrisonburg, Virginia

Design by Michael Collica